From Rebel to Radical Innovator

Leading the Transformation Through Circularity

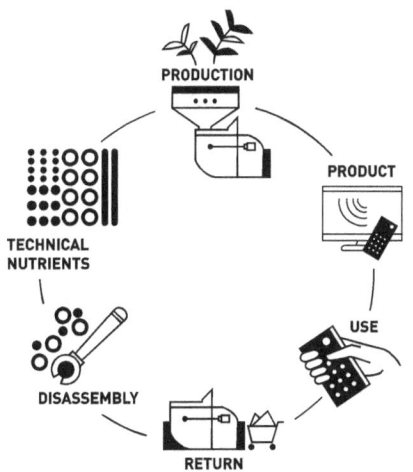

ALBIN KAELIN

Copyright © 2024 Albin Kaelin
Published in the United States by Leaders Press.
www.leaderspress.com

All rights reserved. No part of this book may be reproduced or transmitted in any form or by any means, electronic or mechanical, including photocopying, recording, or by an information storage and retrieval system–except by a reviewer who may quote brief passages in a review to be printed in a magazine or newspaper–without permission in writing from the copyright holder.

All trademarks, service marks, trade names, product names, and logos appearing in this publication are the property of their respective owners.

Cradle to Cradle Certified®, C2C Certified®, and C2C Certified Material Health Certificate™ are trademarks of the Cradle to Cradle Products Innovation Institute (C2CPII). Cradle to Cradle® and C2C® are registered trademarks of MBDC LLC. Cradle to Cradle Design™ is a trademark of MBDC LLC.

The Cradle to Cradle® concept is a biomimetic approach to the design of products and systems. It models human industry on nature processes, where materials are viewed as nutrients circulating in healthy, safe, biological and technical metabolisms.

Given the international copyrights and trademarks related to Cradle to Cradle® the following applications were previously made only in Switzerland for use by epeaswitzerland: Reference Model Cradle to Cradle™, Reference Model™ for Cradle to Cradle (Switzerland and International), Supply Chain Domino Knowledge Transformation™, C2C Vision for a Safe and Circular Future™, Cradle to Cradle DNA Buildings + Facility Management™, Compass Cradle to Cradle™, Network of Trust by epeaswitzerland™, Circular Accounting by epeaswitzerland™, Circular Integrated Company by epeaswitzerland™, Cradle to Cradle for Professionals by epeaswitzerland™, Toolbox Cradle to Cradle™, Transformation by Circular Design Process™, Cradle to Cradle Lighthouse CASE STUDY epeaswitzerland™, and Circular Lighthouse CASE STUDY epeaswitzerland™ are trademarks of epeaswitzerland gmbh. Innovation and Leadership are the Only Survival Strategies™ is a trademark of Albin Kaelin GmbH.

ISBN **978-1-63735-284-7** (pbk)

ISBN **978-1-63735-285-4** (hcv)

ISBN **978-1-63735-286-1** (ebook)

ISBN **978-1-63735-354-7** (audiobook)

Library of Congress Control Number: **2024904769**

ALBIN KAELIN

"I'm actually left-handed. When I got to school, I was forced to write with my right hand. This early encounter with conformity sparked my curiosity and led me to question the status quo. It taught me not to accept anything whose sense I do not understand. My drive to make sense of the world around me has shaped my life and my work. It has led me to champion innovations that transform production concepts, turning waste into 'nutrients' for other products. By challenging the old system and forging new paths, I found my unique role at the intersection of tradition and innovation. This is where I confidently embrace my responsibility to inspire change towards a circular economy."

From 1981 to 2004, Albin Kaelin was Managing Director of Rohner Textil AG in Switzerland. Under his leadership, the company won nineteen international design awards in the 1990s. In 1993, Albin stimulated the development of the product line Climatex® (www.climatex.com) and thus, the first Cradle to Cradle® products worldwide. In 2001, he was awarded the UBS Key Trophy as the "Rhine Valley Entrepreneur of the Year."

- 2002-2007: Member of the Fein Elast Group supervisory board in Austria
- 2005-2009: CEO of EPEA Internationale Umweltforschung GmbH in Hamburg; in 2007, in addition to being CEO of EPEA Netherlands
- 2009-present: Founder, owner, and CEO of epeaswitzerland gmbh
- 2021: Awarded CEO Today Europe Award
- 2022: Won ten awards
    - M&A Today Global Awards, Lawyers International, and Global 100 > Best CEO of the Year
    - CEO Today Europe > Manufacturing–Leader of the Year
    - Business Worldwide Magazine CEO Awards > Most Innovative CEO in the Worldwide Textile Industry and > Business Leadership and Outstanding Contribution to Sustainable Manufacturing

- Acquisition Int. > Best Specialist Cradle to Cradle ® Products & Services Provider
- CIO Views Magazine and Fortunes Crown > Top 10 Most Inspiring CEO in Business in 2022
- World Finance Innovation Awards > Textile Industry epeaswitzerland

• 2023: Won eleven awards
- M&A Today Global Awards, Global 100, Corporate America Today and Lawyers International 100 > Best CEO of the Year
- Acquisition International > 2023's Best Specialist Cradle to Cradle® Products & Services Provider
- Fortunes Crown > The Most Innovative Technology Leaders to Watch in 2023
- The European—Global Sustainability & ESG Awards 2023 > Best CEO in the Textile Manufacturing Industry > Best Cradle-to Cradle Products & Services Innovation Experts > Most Sustainable Textile and Materials Value Creation Leadership
- IE 100 > Most Innovative CEO of the Year - epeaswitzerland gmbh - Europe
- Corporate Livewire > Household Hardware Specialists of the Year

• 2024: Won nine awards
- M&A Today Global Awards, Lawyers International Legal 100, Corporate America Today 2024 and Global 100 > Best CEO of the Year
- Lawyer International's—Legal 100, Corporate Amerika Today 2024—Awards > Best CEO of the Year and Best Specialist Cradle to Cradle ® Products & Services Provider 2024
- IE 100-2024 > Most Innovative CEO of The Year - epeaswitzerland gmbh—Europe
- THE EUROPEAN > Most Innovative CEO in the Textile Manufacturing Industry 2024
- Pioneers in Cradle-to-Cradle Innovation: Leading Experts in Textile 2024 > Trailblazer Award for Circular & Sustainable Textile Innovation 2024

- 2025: Won two awards: Global 100, Corporate America Today > Best CEO of the Year – Albin Kaelin

epeaswitzerland gmbh

epeaswitzerland gmbh supports companies in different areas of activities in the development and implementation of Cradle to Cradle® design concepts. www.epeaswitzerland.com

With an experienced, internationally oriented, and interdisciplinary working management team, Cradle to Cradle® projects are implemented in all industries worldwide.

Acknowledgments

For twenty-five years, I experienced a headwind and was considered a rebel. Now, the time spirit touches the surface, and I am recognized as an innovator. To receive a tailwind makes it move faster with less energy. Remain firm and never give up your principles.

"Innovation and Leadership are the only survival strategies."

"If you compare, you start to compromise."

"This process never ends; if you never get to this point, you never go to survive."

Changing a mindset is challenging and causes many contradictions, discussions, criticism, and disagreement.

I would like to thank my family and friends, especially my wife, Anna, and daughter Sarah. Sarah has become the mother of Ilaria, and I am now a grandfather for the first time. The next generation has now entered my life.

In 1992, I met William McDonough and Michael Braungart, the visionaries of Cradle to Cradle®, introduced by Susan Lyons, the former executive vice president of Design and Marketing, then the President of Designtex, a Steelcase company in the US. I clicked at the very first meeting, and I immediately understood what had to be done. A different understanding of the world has since been revealed to me. Thanks to Susan, Bill, and Michael for their friendship of over thirty years. With their vision, they have created an orientation and the Cradle to Cradle Certified® certification standard that guides many products, companies, and even legislation. I was able to make my contribution to implementing the transformation of the industry in countless real-world examples.

With the Network of Trust by epeaswitzerland™, we were able to be successful. Special thanks to the team of epeaswitzerland gmbh, especially Daniel Aeschbacher for his support with writing the book, our motivating customers, the network of trust of thousands of companies and suppliers worldwide, and other business associates.

Alinka Rutkowska, the CEO of Leaders Press, approached me with the idea of writing a book. This was a push to start a new project to give managers, economists, and the future generation transformation tools for a more livable future. Thank you, Alinka; now the book is published.
- Albin Kaelin

William McDonough, Co-Founder Cradle to Cradle®

"As an architect and designer, I celebrate collaborators and products that meet the usual criteria of aesthetics, cost, and performance and, most critically at this time in the world, integrate exquisite ecological and social wisdom and intelligence. Working with Albin is a privilege because he weaves our ideas and hopes into reality. For me, when I imagine something beautiful, I think of it as possible. Albin, with alacrity and grace, translates vision into reality. He is at the front line of the Cradle to Cradle®-inspired revolution."

Prof. Michael Braungart, Co-Founder of Cradle to Cradle®

"My knight Albin"

Author's Four-Generation Family

Parents:

"You always go your own way and asserted yourself."

"Albin, my son, you are angry, I know you cannot stand incompetence."

Wife Anna:

"For my husband, I am often a critical sparring partner. I question his ideas and decisions in a provocative manner. Ninety-eight percent of the time, his decisions get confirmed. The remaining two percent involve selecting the right golf club or where we go on vacation."

Daughter Sarah:

"Rarely is the conventional way taken but, without exception, a solution is found that you stand behind while facing all the challenging consequences."

Granddaughter Ilaria:

"Opa, give me five!"

From Rebel to Radical Innovator

Do you want to Design your own Product?

Interactive Online (2 hours)

Workshop Cradle to Cradle Design Innovations

CONTENT
Author will offer Interactive Online Workshop Cradle to Cradle Design Innovations, based on the Cooking Book Chapter.

In Working Groups, Participants will become Designers and will create their own Cradle to Cradle, Circularity Products. The Group will present the Product with the Audience, Feedback will be given after all Groups have presented.

INCENTIVE For Free for Consumers who will purchase the Book

EVENTS 08th November 2024: 2 pm CST Asia, 2pm CET Europe, 2 pm EST US
13th December 2024: 2 pm CST Asia, 2pm CET Europe, 2 pm EST US
21st February 2025: 2 pm CST Asia, 2pm CET Europe, 2 pm EST US

Additional Events can be offered depending on demand

LANGUAGE English, optional German, French

PARTICIPANTS Maximum 20 Participants each Workshop (4 groups for 5 persons)

Interested?
Please register at:

Book Announcement:
From Rebel to Radical Innovator,
Leading the Transformation Through Circularity

Author: Albin Kaelin

Table of Contents

Acknowledgments vii

Author's Four-Generation Family ix

Chapter 1: The Chemical Industry 1
    1890s 1
    1900s 2
    1930s 2
    1939-1945–World War II 2
    1950s to 1960s– the Golden Age of Chemistry 3
    1970s to 1980s: the Mature Age 4
    1972: Club of Rome–The Limits of Growth 5
    1973: Oil Crisis 5
    1976: Seveso, Italy Disaster 5
    1984: Bhopal, India Disaster 5
    1986: Schweizerhalle, Switzerland– Greenpeace Activities 6
    1987: EPEA (Environmental Protection Encouragement Agency) . 6
    1990s onwards: the Green Chemistry Age 7
    1992: Hannover Principles 7
    1993: The First Cradle to Cradle® Product 8
    1995: McDonough Braungart Design Chemistry (MBDC) 8
    Green Chemistry 8
    The 21st Century: Certification Schemes and Initiatives 9
    2001-2007: EU Commission-Registration, Evaluation, Authorization, and Restriction of Chemicals (REACH) 9
    2008: California Green Chemistry Law 10
    2010: Cradle to Cradle Products Innovation Institute (C2CPII) 10
    2011: Greenpeace Detox Campaign 11
    1995-2015: the Mature Age in the US and Europe, the Growth Age in Asia 11
    Challenges with the Quality of Chemicals Produced in Asia 13
    Textile and Fashion Industry Focus 14
    2014-2019: Fashion Positive Initiative 14

2017: Fashion for Good 15
2019: VF International, Napapijri "Circular Series" 15
Climate lawsuits against states and companies 16
2020s: Government Initiatives for Circularity 16
2021: EU Commission "The New Green Deal" 17
2021: Frosch Brand's Circular Flexible Packaging Pouch 18
2021: European Chemicals Agency (ECHA) – SCIP Database 18
2022: EU Commission and Due Diligence Supply
Chain Regulation.. 19
Polytetrafluoroethylene (PTFE) and Per- and Polyfluoroalkyl
Substances (PFAS) 20
Today and the Future.. 21

Chapter 2: Transforming Industry 23
Surviving the Impossible... 23
A Customer Initiative 25
Cradle to Cradle® Design Innovations, Rethinking the Way We
Make Things, Climatex® the Worldwide First Cradle to Cradle®
Product 27
A Fire Changes the Game... 29
Crossroads 31
Facelift for the market... 32
Science Fiction 33
Off to New Adventures.. 35
Revolutionizing Resilience: Leading the Chemical Industry
into a Sustainable Future 36

Chapter 3: Transforming Chemistry - Safe and Circular
Chemistry - Case Studies... 39
The Future through Circular Science Cradle to Cradle® 42
Defining Circular Science 45
Innovation in Circular Science Cradle to Cradle®.......................... 49
Toolbox Cradle to Cradle™ Methodology....................................... 52
A Safe and Circular Cradle to Cradle® Future Defined by
Innovation and Sustainability 55

Chapter 4: Transforming Materials – Safe and Circular Materials - Case Studies 57
Transformation from Negative to Positive Definition Case Study 57
Confrontation – Transformation..57
Fighting against Goliath; Case Studies 62
Developing the Business Case; Case Studies 66
Developing High Tech Materials; Case Studies 69

Chapter 5: Revolutionizing Education - Embracing a New Ethos in Industrial Design and Economics..........................73

Chapter 6: Practicing Design Innovations........................81

Chapter 7: Transforming Design with Cradle to Cradle® Principles...93

Chapter 8: Transforming Science 97
Material Health ABC-X Methodology................................97
A Personal Journey Towards Understanding 99
Innovation Through Collaboration 100
Looking Ahead: The Future of Material Science 100

Chapter 9: Cradle to Cradle Certified® Certification 103
From Linear Certification Schemes to Circular for all Products of all Industries 103
Innovations in Material Health and Reutilization 107
Overcoming Barriers to Circular Transformation 108
The Role of Policy and Regulation 108
A Vision for a Circular World 109

Chapter 10: Transforming Products – Safe and Circular Products - Case Studies 111
Textiles..111
Home Textiles 111
Work Wear 116
Fashion 117

Lifestyle Fashion 122
Footwear 125
Paper & Packaging 127
Plastics 133
Cleaning Detergents 138
Wood 140
Construction Industry, Building Materials 144
Ceramics 145
Electronics 148
Flooring 151

Chapter 11: Transforming Systems - From Linear to Circular Systems - Case Studies 157

DNA Buildings + Facility Management, Circular Accounting by epeaswitzerland™ 158
Industrial Composting, Navigating Challenges and Innovations 160
The Circular Economy and the New Green Deal 162
Transforming the Textile Industry in Indonesia: A Journey to Circularity and Cradle to Cradle® 163
Transitioning to Circular Cleaner Production 165
Safe and Circular Products and Closing the Loop 166
Concept Framing for Countries: EU Commission's Green Deal 170
The Challenge of Circularity in the Electronic Industry 173
Consumer Behavior in the Context of Retail Transparency and Circularity 175

Chapter 12: Transforming Management - Management Tools for Circularity - Case Studies 179

Circular Accounting by epeaswitzerland™ 179
Transparency in Products 180
Leadership and Innovation in Sustainability 180
Building the Team: Aligning Individual Values with Circularity Goals 181
Supply Chain Domino-Knowledge Transformation™ 181

Table of Contents

    Management Tools for Circularity..181
    Marketing Based on Science: Avoiding Greenwashing...............182

Chapter 13: Embracing Startups and Their Potential - Case Studies 185
    Start-Ups - Proof of Concept 185
    Beverages Packaging 186
    Cosmetics 188
    Textiles, Fashion 190
    Creating Networks to Leverage Circularity....................................192

Chapter 14: Transparency for Consumers - Case Study 195

Epilogue 199

References / Companies 201

References / Literature 205

About the Author 219

References / Companies

Alfa Klebstoffe . Amer Sports . Salomon . Aquafil . Archroma . Artigo . Asahi Kasei . Avient . Bauwerk . Bayonix . Braungart EPEA Int. . Calida . Circular Clothing . Clariant . Climatex . Cradle to Cradle Product Innovation Institute . DyStar Colours . epeaswitzerland . Fein Elast . Grabher . Heierling . Inogema . Karl Dieckhoff . Knauf Ceilings . Lanz Natur . Laufen Bathroom . Lauffenmühle . Lucart . Luzi . Mary Rose . MBDC . Migros . Mondi . Napapijri . Next Generation . OceanSafe . Pfister Vorhang Service . Kahatex . Rohner Textil . Sanko Tekstil . Sens eRecycling . Siegwerk . Step Zero . Swarovski . Tana Chemie . Tanatex Chemicals . Textilcolor . Trigema . Trimo . TVS . VF Corporate . Voegeli . Werner & Mertz . Windmoeller . Wolford

Abbreviations

3D printing = three-dimensional printing
BASF = Badische Anilin- und Sodafabrik
BAT = Best Available Technologies
BPA = Bisphenol A
C2C® = Cradle to Cradle®
C2CPII = Cradle to Cradle® Products Innovation Institute
CBM = Circular Business Model
CE = Circular Economy
CI = Circular Integration
CMR = carcinogens, mutagens, reprotoxic
CO/PET = cotton/polyester blend
CO_2 = Carbon Dioxide
Cooperative = Circular Clothing Genossenschaft
CSDDD = Corporate Sustainability Due Diligence Directive
DDT = dichloro-diphenyl-trichloroethane
EBIT = Earnings before interest and taxes
ECF = Elemental Chlorine-Free
ECHA = European Chemicals Agency
EPA = U.S. Environmental Protection Agency
EPEA = Environmental Protection Encouragement Agency
EPEA GmbH Part of Drees & Sommer = Environmental Protection Encouragement Agency
EPEA Internationale Umweltforschung GmbH
EPEA Netherland = Environmental Protection Encouragement Agency
epeaswitzerland = Environmental Protection Encouragement Agency
ESG = Environmental, Social and Governance
EU = European Union
EUAP = EU Action Plan on Sustainable Finance
FFG = Österreichische Forschungsförderungsgesellschaft GmbH
FSC = Forest Stewardship Council
GCCA = Global Cement and Concrete Association
GDP = gross domestic product
GHG = Greenhouse gas
GOTS = Global Organic Textile Standard

GPII = Green Products Innovation Institute
GRI = Global Reporting Initiative
HAB = harmful algal bloom
ICMESA = Industrie Chimiche Meda Società Azionaria
ICT = information communication technology
LCA = Life Cycle Assessment
LDF = Low-Density-Fiberboards
LI = Linear Integration
MBDC = McDonough Braungart Design Chemistry
M-Check = Migros Check
MDF = Medium-density fibreboard
MG = Machine glazes paper tissue products
MI = McDonough Innovation
NGO = Non-governmental organization
OEM = Original Equipment Manufacturers
OPEC = Organization of Arab Petroleum Exporting Countries
PBDE = Polybrominated diphenyl ethers
PE = Polyethylene
PET = polyester
PFAS = Per- and Polyfluoroalkyl Substances
PFC = long-chain-perfluorinated chemicals
PFOS = Perfluorooctane
PPM = parts per million
PTFE = Polytetrafluoroethylene
PU = polyurethane
PVC = Polyvinyl chloride
REACH = Registration, Evaluation, Authorisation and Restriction of Chemicals
ROJ = Return of Investment
SABS = Sustainable Accounting Standards Board
SCCP = short-chain chlorinated paraffins
SCIP Database = Substances of Concern In articles as such or in complex objects (Products)
SDGs = UN Sustainable Development Goals
SDS = Safety Data Sheets
SIEFs = Substance Information Exchange Forums
SME = small and medium-sized enterprises
SUPs = single use plastic products

Abbreviations

TCDD = Tetrachlorodibenzodioxin
TCF = Totally Chlorine-Free
UK = United Kingdom
VAT dyes = Water Insoluble Dyes
VF International = VF Corporation
VOC = volatile organic compounds
WEEE Forum = Forum for Waste Electrical and Electronic Equipment
WFD = Waste Framework Directive
WWF = World Wildlife Foundation

CHAPTER 1

The Chemical Industry

It's All About Chemistry

Between ancient times to the present day, the chemical industry has been profoundly transformed. Chemistry has long had a great impact on the world. The craftsmanship of chemistry dates back to 7000 BC when Middle Eastern artisans used alkali and limestone to create glass and sulfur. The creation of glass using alkali and limestone was a significant achievement. However, glass production led to substantial deforestation as wood was the primary fuel for furnaces. In China, the invention of gunpowder, using saltpeter, transformed warfare, causing widespread devastation and altering the course of history.

The First Industrial Revolution and Beyond

The 19th century marked the beginning of the chemical industry as we know it today. The first Industrial Revolution spurred the establishment of large-scale chemical plants in Great Britain (1740), France (1766), Russia (1805), and Germany (1810). The emergence of industrial chemical plants led to environmental pollution and worker exploitation. The textile and glass industries, which drove soda production, were responsible for air and water pollution. The advent of artificial fertilizer plants improved agricultural productivity, but also led to soil degradation and increased reliance on chemical inputs.

1890s

In the latter half of the 19th century, we witnessed remarkable advances in organic chemistry, paving the way for producing synthetic dyes from coal tar and revolutionizing the textile industry. The synthesis of dyes from coal tar, while revolutionizing the textile industry, also introduced harmful environmental pollutants. The mass production of sulfuric acid and electrolytic methods in Germany brought about air and water pollution and health hazards for workers.

In 1899, the textile industry was revolutionized by artificial fibers when Glanzstoff (Germany) introduced rayon fibers (a cellulose fiber from wood). In the 1890s, companies in Germany began the mass production of sulfuric acid and implemented electrolytic methods. Through rapid concentration in the chemical industry, scientific and technological development, patent monopolies, and commercial politics, Germany retained a monopoly on producing organic dyes and intermediates until World War II.

1900s

In 1909, the American Cyanamid Company in the US made significant strides with the introduction of synthetic fertilizers. These new fertilizers significantly increased crop yields but also contributed to soil depletion, water pollution, and reduction in biodiversity. Du Pont (US) introduced nylon in 1928. Developing synthetic fibers like nylon produced non-biodegradable waste and environmental contamination.

This period set the stage for what would later be known as the golden age of the chemical industry.

1930s

By 1930, the United States began to produce plastics and fibers. Polymer science involved manufacturers of plastics, resins, paints, and adhesives, with chemical engineering becoming the driving force of the industry.

1939-1945—World War II

During World War II, synthetic rubber was developed when the war cut off supplies of natural rubber from Asia. While this invention addressed the shortage of natural rubber, it also introduced vast amounts of non-biodegradable materials into the environment. The war-driven advancement in chemicals further led to an increase in hazardous waste and environmental pollution.

After the war, production shifted from coal to petrochemicals derived from oil or gas (fossil fuels). Petrochemicals made from oil became the industry's largest sector. With it came an increased dependency on fossil fuels, contributing to greenhouse gas emissions and global warming.

Now, the trend is away from petrochemicals towards renewable resources derived from nature. This marks a paradigm shift for the energy industry.

1950s to 1960s— the Golden Age of Chemistry

In these transformative years, the industry saw unprecedented growth in the production of synthetic textile fibers, expanding at a staggering rate of twenty percent annually. The magnitude of this growth was akin to the meteoric rise of the software industry in contemporary times, placing companies like Du Pont in a role comparable to that of Microsoft in the digital age.

During this period, agriculture was on the cusp of a revolution. New chemical products with seemingly boundless potential were being discovered and rapidly introduced to the market. I recall how these innovations were swiftly embraced by farmers and the wider public, who were captivated by the apparent miracles unfolding before their eyes. Pesticides, for instance, had shown their efficacy during the tumultuous times of the Second World War, when DDT was used to combat insects that spread malaria and typhus, thereby saving countless lives.

In those early years, the industry was buoyed by optimism and a pioneering spirit, largely unaware of the potential downsides of these chemical marvels. It was a time when the detrimental effects of these substances on material health and the environment were not yet fully comprehended or acknowledged. The historical record shows that an astonishing array of five thousand to six thousand chemical substances were being used in plastic-based products over the past century, with a significant number raising concerns about material health issues.

The regulatory bodies of that era, focused as they were on licensing and distribution, seldom considered the long-term implications these chemicals might have, particularly in relation to their interaction with plastics, water, ultraviolet radiation, or heat—common elements in processes such as textile dyeing or even in everyday appliances like microwave ovens. An intricate balance of four hundred to five hundred substances would be sufficient to fulfill the requirement for each plastic product to ensure compatibility with the biological systems they would eventually interact with.

This reflection highlights a critical aspect of industrial development that transcends the chemical industry: the inherent lag between innovation and a full understanding of environmental and health impacts. It was a pattern all too common in various sectors, where the allure of immediate benefits from new products often overshadowed the potential long-term risks. This narrative serves as a poignant reminder of the delicate balance between progress and sustainability, an issue that industries continue to grapple with as they pursue innovation.

The 1950s, therefore, stand as a testament to the remarkable progress in industrial chemistry and a developing awareness of its potential drawbacks. As we continue to innovate and develop new materials and technologies, it's crucial that we also invest in comprehensive research to fully understand their impacts. Only by doing so can we ensure that today's advancements do not become tomorrow's environmental and health crises.

1970s to 1980s: the Mature Age

The Mature Age of the chemical industry during the 1970s and 1980s was marked by industrial crises, disasters, and protests. These significantly reshaped the industry's approach to safety, environmental impact, and resource management. This period saw the emergence of several key events that challenged existing practices and led to transformative changes in the industry.

1972: Club of Rome—The Limits of Growth

The Club of Rome's report, "The Limits to Growth," highlighted the finite nature of resources and the unsustainable trajectory of industrial growth. It raised concerns about the long-term viability of the prevailing economic models in the Western world. The report emphasized the need to shift towards more sustainable practices in industries, including the chemical sector.

1973: Oil Crisis

The oil crisis of 1973, triggered by the oil embargo by the Organization of Arab Petroleum Exporting Countries (OPEC), had a profound impact on the global economy and the chemical industry. The dramatic rise in oil prices from $3 to nearly $12 per barrel led to increased costs for raw materials in the chemical industry and spurred efforts towards energy efficiency and diversification of energy sources.

1976: Seveso, Italy Disaster

The Seveso disaster was a turning point in the chemical industry's approach to safety and environmental impact. The release of a toxic cloud containing dioxin TCDD from a chemical plant operated by ICMESA, a subsidiary of Givaudan and Hoffmann-La Roche, highlighted the risks associated with chemical manufacturing. This incident led to stricter safety regulations, including the EU's Seveso Directive, which established significant accident prevention policies.

1984: Bhopal, India Disaster

The Bhopal disaster was one of the worst industrial catastrophes in history. The leakage of methyl isocyanate gas and other chemicals from a Union Carbide pesticide plant exposed over 600,000 people to toxic substances, resulting in thousands of deaths and long-term health effects. This tragedy underscored the critical need for enhanced safety measures, risk management, emergency response planning, and corporate accountability in the chemical industry.

1986: Schweizerhalle, Switzerland— Greenpeace Activities

The Sandoz chemical spill in Schweizerhalle was another significant environmental disaster. A fire at the Sandoz warehouse led to the release of toxic chemicals into the Rhine River, causing extensive ecological damage and killing a large portion of the European eel population. The incident sparked protests and actions by environmental groups like Greenpeace, which demanded greater responsibility from the chemical industry for environmental protection.

The Toxics Action Vigil was held at Ciba-Geigy in Basel, Switzerland. Greenpeace and their head of the chemical section, Michael Braungart, held a three-day vigil on 19-21 December 1986, on a chimney of the chemistry company Ciba-Geigy under the slogan: "Only when the last tree is felled, the last river is poisoned, and the last fish is cached, you will notice that you can't eat your money."[1]

1987: EPEA (Environmental Protection Encouragement Agency)

The founding of the EPEA by Michael Braungart, a former Greenpeace activist, represented a shift in environmental advocacy towards proactive cooperation with industry and government. The EPEA aimed to develop methods for evaluating chemicals and promoting sustainable practices within the chemical industry, contrasting the traditional environmental protection approach with a focus on encouragement and positive change.

---

[1] "Toxics Action Vigil at Ciba-Geigy in Basel Switzerland", Greenpeace, Accessed June 4, 2024, https://media.greenpeace.org/C.aspx?VP3=SearchResult_VPage&STID=27MZIFIZYVJH

Collectively, these events highlight a period of intense scrutiny and transformation for the chemical industry. They brought to the forefront issues of resource limitations, environmental degradation, safety lapses, and the need for sustainable development, leading to significant changes in industry practices and regulations.

The first business critiques featured the EPEA's core strength—the ability to work independently and critically with the industry. One such project for Greenpeace was an evaluation of halogenated hydrocarbons. For Ciba-Geigy (now BASF), this project established a method for the evaluation of chemicals.

1990s onwards: the Green Chemistry Age

The 1990s marked a paradigm shift—the beginning of what is now known as the "green chemistry age." This period saw the emergence of the Hannover Principles in 1992, advocating sustainable design. During that decade, we also witnessed the development of the first Cradle to Cradle® products, emphasizing eco-friendly production processes. The late 20th and early 21st centuries marked a significant shift in the chemical industry towards sustainability and environmental consciousness, highlighted by crucial developments such as the Hannover Principles, the inception of Cradle to Cradle® products, and the formalization of green chemistry principles.

1992: Hannover Principles

These principles, developed by William McDonough and Michael Braungart for Expo 2000 in Hannover, advocated sustainable design in architecture and product manufacturing. They emphasize the coexistence of humanity and nature in a sustainable environment, interdependence, and respect for the relationship between spirit and matter. The Hannover Principles also stress the importance of acknowledging the consequences of design decisions on human and ecological well-being, advocating for creating safe, long-term valuable products, eliminating waste, utilizing natural energy flows, recognizing design limitations, and continual improvement through knowledge sharing.

1993: The First Cradle to Cradle® Product

The development of Climatex® upholstery fabrics marked the beginning of the first worldwide products that embodied the Cradle to Cradle® concept—designing products with their entire lifecycle in mind, ensuring they are sustainable, reusable, or safe for biological systems. In 1995, Climatex® Lifecycle products were introduced to the American market by Designtex, a company owned by Steelcase.

1995: McDonough Braungart Design Chemistry (MBDC)

The collaboration between McDonough, an architect, and Braungart, a chemist, led to the establishment of MBDC in Charlottesville, Virginia. This venture was instrumental in promoting and developing the Cradle to Cradle® design framework, which advocates environmentally intelligent design in products and systems.

Green Chemistry

The turn of the millennium saw the formalization of Green Chemistry principles by Paul Anastas and John Warner, who, in 2000, published *Green Chemistry: Theory and Practice*. They laid out twelve principles of green chemistry, setting a framework for the industry to develop more environmentally benign chemicals and processes. In 2001, the Green Chemistry Institute became part of the American Chemical Society, marking a significant step towards mainstream acceptance and implementation of green chemistry principles in the industry. Moreover, the Nobel Prizes in Chemistry in 2001 (to Knowles, Noyori, and Sharpless) and 2005 (to Chauvin, Grubbs, and Schrock) were awarded for research in areas aligning with green chemistry principles, which underscored the importance of this field and its future direction.[2]

---

[2] "2000s", Green Chemistry History, Accessed June 4, 2024, https://www.acs.org/greenchemistry/what-is-green-chemistry/history-of-green-chemistry.html

These milestones represented a paradigm shift in the chemical industry, focusing on sustainability, environmental responsibility, and the health and well-being of people and the planet. Adopting these principles and frameworks has led to significant changes on how chemicals are produced, used, and managed, influencing a broader movement toward sustainability in various industries.

The 2000s saw regulators stepping in to address the growing concerns about chemical safety and environmental impact. Notable developments included the 2005 Texas City refinery explosion, the European Union's REACH regulation in 2007, and the California Green Chemistry Law in 2008. These regulations marked a significant shift towards greater transparency and responsibility in the chemical industry.

The 21st Century: Certification Schemes and Initiatives

The 2010s marked a pivotal era in driving sustainability within the chemical industry, characterized by establishing key certification schemes and initiatives. Notably, the Cradle to Cradle Products Innovation Institute launched in 2010, and the Greenpeace Detox Campaign of 2011, stood as hallmark efforts to foster eco-friendly manufacturing practices and more transparent supply chains. During this period, we also witnessed significant regulatory transformations. These enhanced safety and transparency, mainly through implementing the EU's REACH regulation and the California Green Chemistry Law. Collectively, these initiatives underscored a growing commitment to sustainability. This reflected a crucial shift towards more responsible and environmentally conscious practices in the chemical industry.

2001-2007: EU Commission-Registration, Evaluation, Authorization, and Restriction of Chemicals (REACH)

The European Commission's strategy, initiated in 2001, led to the development of the REACH regulation, which came into force in 2007. REACH was a groundbreaking regulatory framework that demanded greater transparency and control over chemicals in the European Union (EU).

The regulations required registration of chemicals produced or imported in quantities over one ton, evaluation of those made in quantities over one hundred tons, and authorization of substances of high concern, like carcinogens or mutagens or toxic to reproduction (CMRs).

Managed by the European Chemicals Agency, REACH emphasized communication of chemical information throughout the supply chain. This ensured that manufacturers, importers, and their customers were informed about the health and safety aspects of their products.

The formation of Substance Information Exchange Forums (SIEFs) aimed to simplify registration processes and reduce animal testing.

REACH significantly affected retailers who had to provide detailed information about substances in their products, encouraging them to substitute or remove harmful substances.

2008: California Green Chemistry Law

California's Green Chemistry Initiative, established in 2008, focused on reducing the environmental impact of manufacturing processes. The law targeted heavy water use, long transport routes, and wasteful packaging. This involved creating a list of "chemicals of concern" and identifying "priority products" that are heavily used by vulnerable populations. Manufacturers had to certify that their products were free of these harmful chemicals to sell them in California and, in some cases, assess safer alternatives.

2010: Cradle to Cradle Products Innovation Institute (C2CPII)

Opened in California in 2010, the Cradle to Cradle Products Innovation Institute aimed to support and speed up the transformation of products and industries towards more sustainable practices. Accredited by the State of California, the institute globally certifies Cradle to Cradle

Certified® products, promoting sustainable, circular economy principles in product design and manufacturing.

2011: Greenpeace Detox Campaign

Launched in 2011, the Detox campaign by Greenpeace exposed the links between global clothing brands, their suppliers, and toxic water pollution.

Through fieldwork, investigations, and garment testing, Greenpeace released reports revealing the presence of hazardous chemicals in clothing and advocated for a toxic-free fashion industry.

1995-2015: the Mature Age in the US and Europe, the Growth Age in Asia

The years from 1995 to 2015 were marked by contrasting trends in the chemical industry across different regions, with mature markets like the US and Europe experiencing modest growth while emerging markets, particularly in Asia, saw significant expansion.

Europe's chemical industry grew by nearly 60 percent over twenty years from 1995 to 2015. However, this growth was accompanied by a reduction in its global market share, highlighting the relative increase in chemical production in other parts of the world.

In both Western Europe and the United States, the growth rates of the chemical industry aligned closely with GDP. This steady but modest growth maintained the industry's relevance in the general economy, but the rapid expansion in emerging markets overshadowed it.

In the same epoch that witnessed the sustainability initiatives in the West, East Asia, notably China, underwent a dramatic transformation in its chemical industry landscape. The period from 1995 to 2015 was a testament to the burgeoning economic prowess of the Asia-Pacific region, driven by the meteoric rise of the chemical industry. This growth, interwoven with the narratives of industrialization and economic development, painted a dynamic picture of a region ascending to global prominence.

Economic data from this era underscore the significant strides made by the Asia-Pacific region, including countries like South Korea, Singapore, Taiwan, Malaysia, Thailand, and Hong Kong. These nations experienced average gross domestic product (GDP) growth rates two to three times higher than the global average, a phenomenon significantly fueled by the chemical industry. The period saw numerous U.S. and European chemical producers gravitating toward the region, embarking on various projects that represented both opportunity and expansion.

China, a central figure in this narrative, continued to present robust growth opportunities in its chemical sector, even as the country's overall economic growth showed signs of deceleration by 2015. The slowdown to slightly less than 7 percent was a modest dip compared to the 8-10 percent growth rates enjoyed in the previous fifteen years. Despite these shifts, the Chinese chemical industry remained a beacon of growth and potential.

The economic ascent of newly industrialized East Asian countries, including Hong Kong, Singapore, South Korea, and Taiwan, was primarily attributed to capital accumulation. This factor accounted for between 48 to 72 percent of their economic growth, indicating the pivotal role of industrial expansion in these nations' economic narratives.

Furthermore, economic growth and trade openness profoundly influenced the financial sector development in Asia and the Pacific between 1995 and 2011. The region's performance in these aspects notably outstripped other parts of the world, showcasing its burgeoning economic might.

In the late 1990s, the Yangtze Delta region in China witnessed a remarkable growth spurt in chemical industry. New economic support policies that emphasized traditional industrial sectors spurred this development, reflecting a strategic pivot towards bolstering established industries.

All this collectively paints a vivid picture of the chemical industry's significant growth across Asia, particularly in China, driven by a complex interplay of industrialization and economic development from 1995 to 2015. The East Asian tigers' progress provides a window into understanding the dynamics of regional growth, economic policies, and industrial trends that shaped the chemical industry in Asia during this transformative period.

While contributing substantially to the global chemical market, the burgeoning chemical industry in the region faced environmental impact and product quality challenges.

A 2017 report highlighted environmental concerns, revealing that numerous companies in northern China failed to meet environmental standards for controlling air pollution. This issue underscored the ecological costs associated with rapid industrial growth.

Challenges with the Quality of Chemicals Produced in Asia

The quality of chemicals produced in Asia, particularly with respect to impurities and unwanted by-products, remained a critical issue. These quality concerns impacted the entire supply chain, affecting product safety and performance.

Few Asian chemical companies could manage and ensure high-quality standards. This highlighted the need for enhanced quality control and stricter regulatory supervision within the region's chemical industry.

Overall, this period marked the divergent paths of the chemical industry in different parts of the world. While mature markets like the US and Europe maintained steady growth and continued to innovate in environmental and safety standards, emerging markets in Asia, driven by rapid industrialization, faced ecological impact and product quality challenges. These trends underscore the need for balanced growth that incorporates environmental sustainability in economic development.

Textile and Fashion Industry Focus

The latter part of the decade saw initiatives targeting the textile and fashion industry, such as Fashion Positive in 2014, Fashion for Good in 2017, and the Napapijri "Circular Series" by VF International in 2019. These initiatives highlighted the industry's commitment to sustainable fashion and circular economy principles.

2014-2019: Fashion Positive Initiative

The Fashion Positive initiative, launched by the Cradle to Cradle Products Innovation Institute (C2CPII), aimed to transform the fashion industry by promoting sustainable, circular fashion.[3]

Its primary tool was the Material Library, which hosted only Cradle to Cradle Certified® materials, ideally at the GOLD level. This standard indicated the material was optimized according to the Cradle to Cradle Certified® certification criteria.

Fashion Positive was a platform for suppliers, manufacturers, and brands to collaborate and align on sustainable fashion materials.

The initiative focused on incentivizing all stakeholders in the fashion industry, including final users and retailers, to adopt sustainable practices.

A key aspect of Fashion Positive was scalability and the depth of the Material Library—the more comprehensive the library, the greater the impact on the industry.

The initiative also established a collaboration framework with Fashion for Good, another significant sustainable fashion movement. However, Fashion Positive was discontinued in 2019 by the C2CPII.

---

[3] "Trademark Use Guidelines", Cradle to Cradle Products Innovation Institute, Accessed June 3, 2024, https://cdn.c2ccertified.org/resources/certification/policy/POL_Trademark_Use_Guidelines_20200930.pdf

2017: Fashion for Good

Founded by the C&A Foundation with co-founder William McDonough, Fashion for Good brought together a diverse group of stakeholders from the apparel industry, including brands, producers, retailers, suppliers, non-profits, innovators, and funders.

The initiative aimed to reimagine and revolutionize the fashion industry for sustainability, supporting innovations through its Innovation Hub in Amsterdam and a startup accelerator in Silicon Valley.

Fashion for Good recognized the Cradle to Cradle Certified® Product Standard as a measure of sustainable fashion and developed the world's first two Cradle to Cradle Certified® GOLD T-shirts.

A comprehensive How-To Guide was created to assist apparel manufacturers and brands in transitioning toward sustainable practices. The guide was developed with McDonough Innovation (MI), MBDC, and Indian apparel manufacturers.

Fashion for Good collaborated with a consortium of accredited assessment bodies for Cradle to Cradle Certified® certification, MBDC, epeaswitzerland gmbh, and Eco Intelligent Growth, offering services like feasibility studies and certifications for the textile and fashion industry. Working through McDonough Innovation, the consortium was responsible for the publication of the first report.

2019: VF International, Napapijri "Circular Series"

Napapijri's Circular Series represented a significant commitment to sustainable and innovative design. The Skidoo Infinity jacket, developed over three years, was a milestone in employing recyclable materials and circular economy models.

The project was part of a broader effort to innovate the fashion industry's impact on the environment, incorporating a digital take-back program for chemical recycling.

In 2020, supported by epeaswitzerland gmbh, after a year of optimization and innovation, Napapijri aimed to transform its entire supply chain to encompass safe materials, chemicals, and dyes, achieving third-party certification at the Cradle to Cradle Certified® GOLD level with their Circular Series Products.

The project involved a global supply chain of forty-seven suppliers and was documented as a case study by the Ellen McArthur Foundation.

Climate lawsuits against states and companies

Climate legal cases are increasingly being used to compel countries, municipalities, and companies to reduce their carbon emissions or achieve net zero.[4] Courts are finding strong human rights linkages to climate change, leading to extended protection for the most vulnerable in society. Governments and corporations are pressured to act toward climate change goals. The lawsuit cases have more than doubled from 884 in 2017 to 2,180 in 2022, most still in the US, but with more and more cases in developing countries and Small Island Developing Countries.[5]

It will only be a matter of time before states and companies are brought to justice for harming the environment and citizens by wasting resources, using toxic substances in products, and generating microplastics.

2020s: Government Initiatives for Circularity

The 2020s ushered in a new era of environmental consciousness, reflected in a wave of government-led initiatives focusing on circularity

---

[4] "Climate Lawsuits Are On The Rise. This Is What They're Based On.", Columbia Climate School, Accessed June 4, 2024, https://news.climate.columbia.edu/2023/08/09/climate-lawsuits-are-on-the-rise-this-is-what-theyre-based-on/

[5] "Climate litigation more than doubles in five years, now a key tool in delivering climate justice", Columbia Law School | Columbia Climate School | Sabin Center for Climate Change Law, Accessed June 4, 2024, https://climate.law.columbia.edu/news/sabin-center-unep-release-global-climate-litigation-report-2023-status-review

and sustainability. The EU Commission's "New Green Deal," approved in 2021, set ambitious targets for climate neutrality and sustainable economic growth. The launch of the Frosch brand's circular flexible packaging pouch in the same year marked a significant step towards sustainable packaging solutions. Additionally, the European Chemicals Agency (ECHA) introduced the Substances of Concern in Articles or Complex Objects (Products) Database, SCIP Database, in 2021, which tracked substances of concern in products, further emphasizing transparency and safety in the supply chain.

2021: EU Commission "The New Green Deal"

The EU's New Green Deal, approved by the EU Parliament on February 10, 2021, represents a comprehensive strategy to transform the EU into a climate-neutral continent by 2050. It is structured around three main objectives.

1. Climate Neutrality by 2050: The Green Deal aims to make Europe the world's first climate-neutral continent by 2050, a goal that presents both a significant challenge and opportunity. It involves ambitious measures to reduce emissions drastically across various sectors, including energy, transportation, agriculture, buildings, and critical industries like steel, cement, information communication technology (ICT), textiles, and chemicals.
2. Transition to Circular Economy: Transforming from a linear to a circular economy is central to the Green Deal. This transition focuses on sustainability and efficiency, where resources are reused, and waste is minimized. It promotes sustainable green transitions across industries and encourages investments in green technologies, fostering new sustainable solutions and businesses.
3. Toxic-Free Environment: Ensuring a toxic-free environment is critical to the Green Deal. It involves setting stricter controls on chemicals and materials and ensuring that the environment is protected from hazardous substances, thereby promoting the health and well-being of European citizens.

The success of the Green Deal relies heavily on the involvement and commitment of the public and all stakeholders. It aims to ensure a just and socially fair transition that leaves no individual or region behind.

The EU tax policies "taxonomy" is a fundamental element of the EU's sustainable finance framework, serving as a vital tool for enhancing market transparency. Its primary function is to guide investments towards those economic activities that are crucial for the transition goals of the European Green Deal. This tax infrastructure operates as a categorization system, establishing criteria for economic activities that contribute to achieving a net-zero emissions target by 2050 and addresses a range of environmental objectives beyond climate change. As an integral part of the EU Action Plan on Sustainable Finance (EU AP), EU tax incentives "taxonomy" play a significant role in redirecting capital flows into sustainable investments. The tax system also aids in managing the financial risks associated with climate change, environmental degradation, and social challenges while promoting overall transparency in the financial sector.

2021: Frosch Brand's Circular Flexible Packaging Pouch

In 2021, Frosch launched a pioneering circular flexible packaging pouch. This innovation represents a significant step in applying circular economy principles in product packaging, emphasizing sustainability, recyclability, and reintegration into the supply chain.

2021: European Chemicals Agency (ECHA) – SCIP Database

Inaugurated in 2021 under the Waste Framework Directive (WFD)–specifically Directive (EU) 2018/851– the SCIP database serves as a pivotal tool for managing data on Substances of Concern in Articles or Complex Objects (Products). This database is crucial in enhancing transparency for hazardous substances in articles and products. Its primary objective is to support the circular economy by streamlining waste management and recycling processes.

From January 5, 2021, companies operating in the EU market and supplying articles falling under the Candidate List were mandated to submit pertinent information to the European Chemicals Agency (ECHA). The SCIP database is instrumental in ensuring the accessibility of this information across the entire lifecycle of products and materials, including at the waste stage, thereby facilitating waste operators and consumers.

Regarded as an ambitious initiative, the SCIP database involves collaboration across multiple EU, national, and global tiers. Its implementation was a strategic move to provide comprehensive data on waste composition, enhancing reusability or recyclability. The database became operational in January 2021 and is a directive under the EU Waste Framework Directive.

The legal framework and informational requisites of the SCIP database, as outlined under Directive (EU) 2018/851, has sparked significant discussion. This discourse delves into the impact of the database on various industries, authorities, and NGOs and the potential challenges posed, including exceeding legal limits or necessitating additional data collection from global supply chains.

The SCIP database is pivotal in promoting greater transparency about hazardous substances in products and waste. This step is crucial for optimizing recycling and waste management practices, aligning with the EU's environmental objectives.

2022: EU Commission and Due Diligence Supply Chain Regulation

In 2022, the EU Commission implemented regulations for due diligence in supply chains, focusing on human rights and environmental compliance. This initiative represents an approach to corporate responsibility, ensuring that the industry's growth does not come at the expense of ethical and ecological standards. The approval by the EU Parliament Corporate Sustainability Due Diligence Directive (CSDDD) was agreed on 15 March 2024.

This due diligence supply chain regulation mandates businesses in EU member states to ensure compliance with human rights and environmental legislations within their supply chains. Companies must conduct risk analyses to provide transparency and legal compliance across their global supply chains. This regulation highlights companies' challenges in understanding their supply chain beyond the first tier.

These initiatives and regulations collectively represent the EU's comprehensive approach to addressing climate change, promoting circular economy principles, ensuring a toxic-free environment, and enhancing supply chain transparency and sustainability.

Polytetrafluoroethylene (PTFE) and Per- and Polyfluoroalkyl Substances (PFAS)

PTFEs and PFAS are under scrutiny for potential bans in the US and the EU, signaling a move towards a more toxic-free environment. PTFE, a subgroup within the broader category of fluorinated polymers, is a member of the PFAS family, accounting for a significant portion of the market.

PFAS, a vast group of chemicals, has accumulated in human and wildlife populations over time. Alarmingly, PFAS has been detected in the blood of 99 percent of Americans, a statistic likely mirrored globally. These substances can cross the human placenta, resulting in newborns already carrying industrial chemicals.

PTFE, commonly recognized in its application as Teflon by DuPont, possesses unique characteristics, including excellent heat resistance, electrical insulating properties, and high-water repellence. Its non-stick nature makes it a popular choice for coatings on kitchen appliances like baking sheets. Additionally, PTFE is used in membranes found in outdoor jackets and other textile products.

In a significant regulatory move, the U.S. Environmental Protection Agency (EPA) outlined a plan in October 2021 to address PFAS concerns. This initiative was further advanced with the launch of an interactive PFAS Analytic web page on 11 January 2023. Similarly, in January 2023, a collective proposal by Germany, Denmark, the Netherlands, Norway, and Sweden was submitted to ECHA to restrict a wide array of PFAS under the REACH regulation. This proposal, targeting the restriction and potential ban of approximately ten thousand PFAS substances, is now publicly accessible on ECHA's website, marking a critical step towards eliminating this entire group of chemicals for a safer environment.

Today and the Future

As we look to the future, the chemical industry stands at a crossroads. Environmental sustainability, ethical production, and safety challenges drive innovation and transformation. The industry's response to these challenges will shape its path forward, focusing on green chemistry, circular economy principles, and responsible manufacturing practices. After all the research advancements in green chemistry and engineering, mainstream chemical businesses must fully embrace the new technology.[6]

Currently, more than 98 percent of all organic chemicals originate from petroleum. However, green chemists and engineers actively try to ensure that their research and innovations from the laboratory environment influence corporate decision-making. This crucial shift aims to pave the way for a future that is not only bio-based and safe, but which also adheres to the principles of a circular economy.

Advancements in technology and increased collaboration among stakeholders are essential in addressing the industry's complex challenges. Digitalization, artificial intelligence, and biotechnology are emerging as critical drivers of sustainable innovation, offering new opportunities for eco-friendly and efficient processes.

The chemical industry's journey from its ancient origins to the present reflects a continual evolution driven by societal needs, technological advancements, and environmental awareness. As the industry moves forward, its ability to adapt and innovate while maintaining a commitment to sustainability and responsibility will be crucial in shaping a greener, more sustainable future for all.

---

[6] "Today and the future", Green Chemistry History, Accessed June 4, 2024, https://www.acs.org/greenchemistry/what-is-green-chemistry/history-of-green-chemistry.html

CHAPTER 2

Transforming Industry

As I reflect on the journey of Rohner Textil AG and my path within industries, I remember the profound transformations we undertook, driven by a vision beyond mere profit or traditional manufacturing paradigms. This chapter is a story of resilience, innovation, and a relentless pursuit of sustainability, unfolding through the adoption of the Cradle to Cradle® design philosophy.

Once upon a time, in Switzerland, there was a tiny textile mill struggling to survive the impossible. No fairy tale dream came true for the company, and it had an abrupt, unwanted, unfriendly end, but not for Albin Kaelin, the former CEO, who marched on unwaveringly, making the dream become reality forty years later.

Surviving the Impossible

The 1990s were a defining era, marked by immense challenges for the European textile industry, including our company, Rohner Textil AG. Economic downturns, escalating environmental concerns, and rapidly evolving market dynamics pushed traditional manufacturing to the brink. As the CEO, I was at the helm during these turbulent times, realizing that our existing business models and practices were increasingly unsustainable. It was clear that radical change was essential for our survival.

Our mill was nestled south of Lake Constance, Europe's largest freshwater reservoir that quenches the thirst of 4.5 million people daily, just a stone's throw from the Rhine River, near the Austrian border and about twenty kilometers from Germany. That location posed a unique challenge: complying with varying environmental regulations across different countries.

That compliance was even more difficult because our historical site building wasn't allowed to undergo structural changes or expansions. It was a challenging environment for any company. Yet we at Rohner Textil AG stood our ground. Our high-tech jacquard weaving and yarn dyeing facilities were our pride. Despite the strong Swiss franc, high employee wages squeezing us from both sides, and stringent environmental regulations, we persevered.

In this climate of uncertainty and desperation, I discovered the Cradle to Cradle® concept, introduced by William McDonough. This wasn't just another business strategy—it was a revolutionary approach to manufacturing. It compelled us to reevaluate every facet of our production and product design process. Rooted in sustainability and both eco-efficiency and eco-effectiveness, Cradle to Cradle® proposed a transformative idea: treating waste not as a Cradle to Grave product but as a beginning, a valuable resource – an asset rather than a liability.

Embracing Cradle to Cradle® was a monumental task, requiring a complete paradigm shift. Initially, there was tangible skepticism within our company to this innovative approach. Questions arose about feasibility, economic viability, and practical implementation. However, amidst these concerns, a realization dawned on us. This radical concept might be the key to weathering the storm and forging a sustainable path forward in manufacturing.

As a team, we began seeing the potential in this new thinking. It was more than just adapting our manufacturing processes—it was about reimagining the entire lifecycle of our products. By integrating Cradle to Cradle®, we aimed to transform challenges into opportunities, redefine the role of waste to nutrients in our production, and position ourselves as pioneers in sustainable manufacturing. It was a journey filled with trials, errors, and triumphs, but one that held the promise of a more sustainable and resilient future for Rohner Textil AG.

How did we manage? Our forward-looking business strategies were our lifeline. We believed that, by innovatively reconciling economic and ecological goals, we could carve out a niche in the market. This approach set us apart and shielded us from being overrun by industry giants.

We weren't just a textile mill: we were innovators in weaving and yarn dyeing, specializing in high-quality upholstery textiles. We produced fabrics for world-renowned brands in office furniture and international textile merchandisers. Our team in Balgach/Heerbrugg in Switzerland blended creativity, technology, ecology, and quality. This combination gave us a competitive edge in this demanding market segment. We were proud to be at the forefront of fashioning the future of textiles, balancing tradition with modernity and quality with sustainability.

A Customer Initiative

In the fall of 1992, a pivotal moment came in my journey as CEO of Rohner Textil AG. Susan Lyons, then the Executive Vice President of Design & Marketing and many years later president of Designtex US (a Steelcase company), reached out to me. She proposed a visit by the renowned US green architect William McDonough, who TIME magazine would, in 1999, name a "Hero for the Planet."

Excited, I agreed and collected Bill from Zurich airport. During our drive to the mill, he uttered three words that would redefine our entire operation: "Waste equals Food." It clicked instantly—a revelation illuminating our mill's path forward. We needed to transition from the linear "Cradle to Grave" model to a circular "Cradle to Cradle®" approach.

At that time, our textile mill, like many global companies, was stuck in the Cradle to Grave mindset, a dead-end approach with no sustainable future. Our yarn dyeing production generated toxic wastewater, which we directed to the public wastewater treatment plant. The authorities demanded we invest in a neutralization and retention basin, but funds for such end-of-pipe technologies were scarce. Moreover, our weaving department produced waste selvages, accounting for up to 13 percent of the fabric weight, which had to be incinerated. The fabric's composition, a mix of cellulosic and non-biodegradable materials, was fundamentally unsuited for the "Waste equals Food" concept. The dilemma seemed insurmountable.

However, in the spring of 1993, our journey towards transformation gained momentum when William McDonough, through DesignTex, arranged for Prof. Michael Braungart, a chemist, former Greenpeace activist, and owner of EPEA Internationale Umweltforschung GmbH, to visit our mill. Initially, there was apprehension about letting a former Greenpeace activist into our company. Would he protest or expose us in the media? But those worries quickly dissipated. The meeting was constructive and visionary.

The first step was to send a request to our chemical and color suppliers asking them to disclose the ingredients of their products for a scientific assessment of their non-toxicity to humans and the environment. All but one declined. The only company willing to cooperate was CIBA-Geigy in Switzerland, later known as CIBA and split into BASF and Huntsman and most recently acquired by Archroma. Of the sixteen hundred color pigments we used, only sixteen (1 percent) were suitable for our project. This revelation showed why other chemical companies were unwilling to disclose their information. The challenge was daunting. How could we market this new sustainable approach when only 1 percent of our color dyes were approved?

Yet this challenge propelled us into a new era of textile manufacturing that was not only environmentally responsible, but also paved the way for innovative and sustainable practices in the industry.

Our collaboration with McDonough marked a transformative era. We set out to revolutionize our manufacturing process, anchoring it in the tenets of sustainability, efficiency and effectiveness. This shift transcended mere operational changes—it was a profound rethinking of our production philosophy. Despite facing skepticism and numerous hurdles, forging a new path in sustainable manufacturing was invigorating.

During this transformative phase, we initiated the reengineering of our production lines to conform to Cradle to Cradle Design™ principles. This process involved a meticulous examination of every aspect of our manufacturing, ranging from the procurement of raw materials to the management of waste products. Our pursuit of materials that met our enhanced sustainability standards frequently drove us toward innovation and developing new solutions.

This era was characterized by intense learning, experimentation, and incremental progress. We encountered a spectrum of challenges, encompassing both technical and financial hurdles. Each obstacle, however, provided an opportunity to innovate, revise our approaches, and advance toward our objective. The journey was rigorous yet empowering as we began witnessing the concrete benefits of our endeavors. Our products evolved to become more sustainable, gaining traction in a market increasingly attuned to environmental concerns.

Cradle to Cradle® Design Innovations, Rethinking the Way We Make Things, Climatex® the Worldwide First Cradle to Cradle® Product

Our collaboration with Prof. Braungart's scientific institute, Environmental Protection Encouragement Agency (EPEA), was a turning point in our journey at Rohner Textil AG. It began a transformative phase, where we committed to producing non-toxic and environmentally friendly textiles. This shift required a fundamental change in mindset, even before technological innovation.

The support and scientific expertise of EPEA were invaluable in realizing our Cradle to Cradle® vision. Their staff guided us through a complete overhaul of our production processes. We reexamined every aspect of our operation—from the materials we used to our methods of processing, dyeing, and finishing textiles. We adjusted to meet the highest environmental standards.

This transformative journey was not just about technical changes—it represented a profound cultural revolution within our organization. We were fostering a new paradigm where sustainability was held in equal esteem to profitability among our stakeholders—owners, employees, suppliers, and customers. This cultural shift emerged as our most significant triumph, establishing a solid foundation for a sustainable future in the industry.

In reflecting on this journey, I recall a crucial moment with Prof. Braungart in the late 1980s. As a Greenpeace activist, his earlier protests against chemical companies, including an incident at CIBA-Geigy, opened the door to an unexpected yet vital dialogue and partnership. This relationship proved instrumental later when we sought CIBA-Geigy's collaboration on our sustainable textile project. Their readiness to share confidential information underscored the deep trust cultivated over the years, bridging past activism and present cooperation.

The fruits of this transformation were immediately evident in the enhanced quality of our wastewater. An independent laboratory's analysis brought to light a startling revelation: our treated effluent mirrored the properties of Swiss drinking water. This groundbreaking achievement eliminated the need for further investment in wastewater treatment. Declaring our materials safe and non-toxic, we achieved compliance and initiated a significant shift in industry paradigms.

However, ensuring our products were safe and adhered to Cradle to Cradle® principles posed another challenge. Our Climatex® upholstery fabrics, patented in 1987 with a climate control seating function, was made of polyester (for the technical cycle) and a blend of wool and ramie fibers (for the biological cycle). This combination hindered circularity, prompting us to redesign the product.

After much debate and trial, we developed and patented Climatex® Lifecycle™. Replacing polyester with wool in the warp made the product biologically cycle-compliant and environmentally friendly. This innovation was launched in 1995 by Designtex at the Guggenheim Museum in New York. Our success spurred competition, but we distinguished ourselves with our commitment to using environmentally friendly dyes and chemicals.

We created mulch for local gardeners from our weaving production waste, providing an ideal fertilizer—embracing the old saying, "Nails and hair fertilize your ground for seven years." Over time, Rohner Textil AG and the Climatex® products garnered nineteen awards, a testament to our commitment to transforming products into safe, circular solutions.

In 1995, McDonough and Braungart furthered their commitment to sustainable design by founding MBDC LLC (McDonough Braungart Design Chemistry) in Charlottesville, Virginia. This partnership marked yet another step in the journey toward a sustainable future in manufacturing.

A Fire Changes the Game

The Dusseldorf Airport fire in 1996 was a watershed moment for us at Rohner Textil AG, particularly for our Climatex® Lifecycle™ products. This tragic fire caused the release of toxic chemicals, dioxin, and chlorides due to the use of PVC (vinyl), particularly in power cables and illegally installed combustion insulation materials. The fire, which caused loss of lives, resulted in stringent new fire safety regulations, which significantly impacted industry and building construction. For us, it meant our products no longer met German requirements, posing a substantial challenge. The disaster reshaped the industry's regulatory landscape, limiting our products' use in larger-volume contract businesses across Europe.

Faced with this challenge, we refused to bury our heads in the sand. "Leadership and innovation are the only survival strategies™," I often said. We embarked on a rigorous journey to find sustainable but effective flame-retardant solutions. The quest was about compliance and upholding our commitment to Cradle to Cradle® principles. We collaborated with the scientific institute, EPEA, and other material science experts and dived into extensive research and development.

After three years of relentless work, we made a breakthrough in 1999 with Climatex® LifeguardFR™. This innovation astounded the testing institutes with its performance—no smoke, no smell, yet passing all tests with flying colors. It was a technical success and evidence of our commitment to sustainability and safety.

At the same time these changes were occurring, new business opportunities began to surface, especially within the transportation and mobility market, encompassing automotive, trains, and airplanes. It's crucial to highlight a novel approach we pioneered during this period. Initially hesitant to enter these highly regulated markets, our company innovated a new business model—we decided to sell licenses to established players within these sectors. This strategic decision led to the successful establishment of three licensing agreements in Canada, Brazil, and Switzerland, marking a significant shift in our business approach.

However, just two months after finalizing the license agreement with the Swiss company, an unexpected turn of events unfolded. Rohner Textil AG was purchased in what appeared for me to be an "unbalanced takeover," a development that took place without my foreknowledge, even my position as CEO.

Looking back on these events, I realize that one of my biggest mistakes was waiting too long to exit. Despite our team's success and substantial financial gains, cultural mismatches and decision-making delays under the new ownership were challenging. Despite receiving a BBB rating from UBS Switzerland and accolades like the Entrepreneur of the Year, the takeover shifted the company's direction.

This experience taught me that companies driven by market share often suppress innovation, fearing it opens doors for competitors. Such companies closely monitor the market for innovations to acquire and frequently shelve them, preventing broader industry progress. In our case, Airbus approached the new owner for innovative materials for their Airbus A380, the world's largest passenger jet. Our Climatex® LifeguardFR™ fabric was chosen for the economy class of this advanced aircraft. However, due to limitations in color matching for the business class, they could not be selected. The new owner's reluctance to produce these innovative fabrics themselves, yet they were initially manufactured at Rohner Textil AG.

I left Rohner Textil AG in the fall of 2004. The company closed in 2008 and was liquidated in 2015. Another Swiss textile company acquired the Climatex® patents and technology, leading to the formation of Climatex AG. This new entity developed Climatex® Dualcycle™, continuing the legacy of innovation and sustainability. The products we initially created, including Climatex® LifeguardFR™, remain in the market, serving residential, contract, and transportation sectors. Our journey, marked by challenges, failures, and triumphs, demonstrates the power of innovation and the importance of sustainability in business, but without any guarantees of corporate survival.

Crossroads

As one chapter of my life was closing, another was beckoning. Amidst the personal and professional upheaval, a new opportunity emerged: Prof. Michael Braungart offered me the position of CEO at his institute. He recognized that, while the scientific foundations of the Cradle to Cradle® approach were well-established, implementing this concept within the industry was still a challenge. Braungart needed someone with industry experience to bridge this gap and drive transformation.

The decision was not straightforward for me, however. First, my background was purely management and industrial, with no formal scientific training. Secondly, the institute was in Hamburg, Germany, a significant distance from my home in Switzerland. Thirdly, the compensation offered was lower than industry standards. Lastly, I was navigating through a personal challenge—divorce. Life often presents us with crossroads, and this was one of those defining moments.

After much contemplation in the fall of 2005, I embraced this new path. I accepted the role of co-CEO for business operations at EPEA Int. Umweltforschung GmbH, with Michael Braungart serving as co-CEO for science. This marked a significant shift in my career from industry to the forefront of environmental research and consultancy.

Another pivotal figure in this new chapter was Peter Donath, who had led the Environment, Health, and Safety department at Ciba Specialty Chemicals from 1996 to 2004. After retiring from Ciba, Peter offered his expertise and assistance without compensation. His extensive experience in the chemical industry was invaluable in implementing the Cradle to Cradle® scientific methodology for circularity within various sectors. Peter and I formed a close bond, united by our shared vision and commitment to sustainable industry practices.

This transition heralded a new chapter, both personally and professionally, characterized by a commitment to sustainability and circular economy principles. It represented a career shift and a contribution to a broader movement aimed at redefining industry practices in harmony with the environment.

Facelift for the market

Introducing Cradle to Cradle® quality was a game-changer, shifting market perceptions about sustainability. It challenged existing norms and established a new benchmark for environmental responsibility in manufacturing and products. It wasn't just about creating a new logo or slogan: it encapsulated the philosophy and approach to sustainability. We faced numerous challenges communicating this vision to a diverse market, including skeptical industry peers and environmentally conscious consumers. This section dives into the marketing initiatives, educational campaigns, and partnerships we forged to elevate our product's brand and its principles.

My first task was to give the science a face that the industry could understand. Cradle to Cradle® became that image. The second challenge was implementing the scientific methodology for circularity and inserting Cradle to Cradle® into an industry that was still operating on linear science-based concepts. Peter Donath, with his global network of chemists and toxicologists, investigated the feasibility of this methodology. The shift in mindset faced loud criticism. For instance, lobbyists from the Life Cycle Assessment (LCA) methodology classified Cradle to Cradle® as "wrong and promoting consumption."

In addressing these criticisms, we had to consider how to encourage industries to disclose confidential, proprietary information to third parties. I defined the institute's role as a "knowledge trustee," leveraging my Swiss background for credibility and my management experience to ensure information security. Communicating the institute's scientific position, aligning with Prof. Braungart and the scientific team, required coherent and comprehensive science-based statements. Although I lacked a formal scientific background, Peter effectively steered this process, earning recognition and credibility among the global scientific community.

A crucial task was to align the scientific views of EPEA Internationale Umweltforschung GmbH in Germany with those of MBDC LLC in the U.S. In 2006, Ken Alston, CEO of MBDC, invited Dr. Christoph Semisch, a senior scientist and chemist from EPEA, and me to discuss the Cradle to Cradle® certification scheme. Initially, fitting the Cradle to Cradle® mindset into a certification standard scheme seemed counterintuitive, challenging, and narrowing. However, after William McDonough and senior scientist Jay Bolus of MBDC presented an initial draft of a possible Cradle to Cradle Certified® Certification, we saw it as an opportunity.

The bold idea was to develop a product certification scheme covering material health, circularity, energy, water, and social fairness for all products across industries. A primary concern was that MBDC and EPEA were certifying their own project products. After an initial phase, we needed to transfer this certification to an independent, not-for-profit, third-party entity. That was risky, as it involved handing over control to an organization that would make, decide, and enforce its interpretation. However, to maintain credibility, independence, and transparency, this was a necessary step. The decision ultimately lay with the owners of MBDC, McDonough and Braungart. This move was a leap into uncharted territory, but it was essential for ensuring the integrity and credibility of the Cradle to Cradle Certified® certification.

Science Fiction

Our journey in introducing sustainable practices to different markets was met with various responses, from enthusiastic acceptance to cautious skepticism. We employed multiple strategies to address these differing attitudes, forged instrumental partnerships for global outreach, and tackled challenges by adapting our approach to suit diverse regulatory and cultural contexts.

In the end, the global response to the Cradle to Cradle® concept, particularly its rapid adoption in the Netherlands, was both a validation of our efforts and a source of inspiration.

The documentary *An Inconvenient Truth* by Al Gore had a profound impact, particularly in the Netherlands, leading to a surge of awareness and action against climate change. This set the stage for another documentary in 2007, *Waste = Food* by Rob van Hattum, which focused on the Cradle to Cradle® design concept. This film showcased our work at Rohner Textil AG, along with a German clothing manufacturer, the Nike EU headquarters, a U.S. furniture manufacturer, and the Ford Motor Company. The film helped catalyze a transformational movement in the Netherlands, eventually leading to the adoption of the "circular economy" concept and formulation of the "New Green Deal" in 2021.

The Netherlands' commitment to renewing itself based on Cradle to Cradle® principles presented a significant opportunity. In response, McDonough and Braungart engaged in various initiatives in the Netherlands, including architectural consulting and public speaking engagements. In 2007, EPEA Int. Umweltforschung GmbH established EPEA Nederland B.V. in Venlo, with me as its CEO. We aimed to support Cradle to Cradle® initiatives locally and create a more significant impact through a team of scientists and consultants.

However, the journey could have been smoother. We encountered competition and skepticism from other consultants, along with difficulties in recruiting scientists. To address this, we created a network of Dutch consulting firms to share Cradle to Cradle® knowledge, leading to a broader concept dissemination. Partnerships with the Chamber of Commerce in Limburg and initiatives like the "First Step to Cradle to Cradle®" voucher program helped propel companies like Van Houtum Toilet Paper (now WEPA) and Mosa Tiles to become beacons in Cradle to Cradle® implementation.

We faced numerous challenges, from defining viable recycled toilet paper to designing compliant building materials and PVC-free carpets. These efforts were groundbreaking and constantly evolving.

Meanwhile, business in Hamburg flourished, positioning us as a global leader in Cradle to Cradle® excellence. The strategy proved financially viable and contributed to our rapid growth. During this period, I traveled extensively, balancing the intense demands of leading two scientific teams, managing finances, supervising projects, and more.

This role was as challenging as it was exciting. It required a versatile skill set, from team management to strategic positioning to negotiating with potential clients and managing growth. Although I learned and discovered much on the job, its intensity and administrative demands made me question whether it was sustainable in the long term. Nonetheless, the journey, filled with learning and innovation, reflected the dynamic and transformative nature of the Cradle to Cradle® philosophy and its impact on the world stage.

Off to New Adventures

My journey didn't end with my tenure at Rohner Textil AG; rather, it began an expansive new chapter. In the fall of 2009, Prof. Braungart and I decided it was time for me to forge my own path. This led to the creation of epeaswitzerland gmbh, a company based in Switzerland, wholly owned by me, with a global reach. Leaving behind my CEO roles in Germany and the Netherlands, I embarked on this new adventure, laser-focused on innovation projects from Cradle to Cradle®.

The evolving sustainability landscape champions adaptability and resilience. The journey of advocating for a sustainable future is ongoing, filled with learning, failures, growth, and the relentless pursuit of innovation.

In 2010, the Green Products Innovation Institute (GPII) (later known as the Cradle to Cradle® Products Innovation Institute or C2CPII) was founded. This initiative, supported by SAP and launched at Google HQ with California Governor Arnold Schwarzenegger, William McDonough, Wendy Schmidt, and numerous other sustainability leaders, marked a significant milestone. McDonough announced the donation of years of intellectual property to the GPII, laying the groundwork for new open standards around the Cradle to Cradle® Design framework.

Fast forward to 2019 when Braungart established a new venture in collaboration with Drees & Sommer, leading consultants in the building industry, forming EPEA GmbH. Meanwhile, epeaswitzerland gmbh was thriving, becoming an accredited assessor for the Cradle to Cradle Certified® certification in 2013 until 2024 and growing into a network of twenty freelance partners, all with a business or industry background, implementing Cradle to Cradle® Design innovations globally.

This focus on innovation began to yield significant results by 2020. Six projects from epeaswitzerland were nominated for the German Sustainability Award (along with companies like Aquafil, Mary Rose, Trigema, Frosch, Wolford, and OceanSafe), with two winning awards (OceanSafe and Frosch). In 2021, the Frosch flexible pouch received the German Packaging Award. In 2023, five of our projects were nominated for the German Sustainability Award, and one received it. My accolades included ten Business Awards in 2022, eleven more in 2023, and nine in 2024 and two in 2025 by September 2024.

These accomplishments weren't just a continuation of my work at Rohner Textil AG—instead, they represented a broader mission to weave sustainability into various facets of manufacturing and product design. The challenges and successes of introducing Cradle to Cradle® concepts to new industries, the collaborations with other companies and innovators, and the ongoing advocacy for sustainable practices were all part of this overarching goal.

Revolutionizing Resilience: Leading the Chemical Industry into a Sustainable Future

"Transforming Industry" transcends the boundaries of a personal account or the growth of a single company. It embodies the story of the chemical industry at a pivotal point of transformation. This narrative intertwines the history of the industry's evolution with modern sustainability strides, illustrating the mutual influence of these two dimensions.

This demonstrates how a single company's innovation can set new benchmarks for an entire industry and the expansive potential of an innovative idea to grow into a movement that surpasses geographical and sector boundaries. It stands as robust testament to the compatibility of sustainability and profitability, challenging the notion that these ideals are in opposition. It underscores how sustainability not only aligns with, but can also bolster, business success.

I hope this story ignites inspiration within the industry and beyond to embrace innovation, highlighting the immense potential of sustainable practices. It is a narrative of ongoing transformation, where each step, big or small, contributes to a future that is more sustainable, responsible, and full of promise.

CHAPTER 3

Transforming Chemistry - Safe and Circular Chemistry - Case Studies

From Linear to Circular: Manufacturing Models

My early experiences at Rohner Textil AG embodied the prevailing linear manufacturing model, where resources flowed straight from extraction to disposal. While this approach catalyzed immense economic growth, it came at a grave ecological cost.

However, the 1990s proved a pivotal turning point as the concept of Cradle to Cradle® Design emerged, challenging linear models. We began realizing that, rather than struggling against Nature, manufacturing could harmoniously coexist within nature's cycles.

This birthed the concept of a circular economy, where waste is designed to nourish new products instead of ending up in landfills or being incinerated. The goal was to eliminate waste by ensuring all materials continually circulate. Unlike conventional recycling, the quality level of the raw materials remains constant throughout multiple product lifecycles, and only "assessed safe" chemicals are used.

While radical, these circular Cradle to Cradle® principles immediately resonated with me, compelling a reimagination of every facet of production. The systemic shift was complex, often met with skepticism, but immensely enlightening. (A more in-depth debate and explanation is provided in Chapter 8: Transforming Science.)

Circular Science Cradle to Cradle®: The Pathway to a Sustainable Future

Circular Science Cradle to Cradle® is an innovative and multidisciplinary approach that integrates sustainability, recycling, worst-case exposure, renewable energy, soil and water, social fairness, and eco-design principles. Cradle to Cradle® is the scientific backbone of the circular economy, in which every product is designed and used to maximize its lifecycle, minimize, or eliminate waste, and ensure that resources are non-toxic and reused in a closed-loop system. It contrasts sharply with the traditional linear 'take, make, waste' model.

Linearity is engraved in industry, which is still primarily based on a traditional and conservative culture of production. The chemical industry includes another dimension of complexity—risk management. This makes it even harder to institute any transformation to a new way of thinking and implementation. The process can take years or even decades to complete. Next to economic feasibility, market demand requires reinforcement through approval by management, as well as encapsulating trust, respect, patience, confidentiality, support, protection, and courage. A rule in business management terminology is that a return on investment should be seen in less than 1.5 years (ROJ), but this is almost impossible in a transformation context. This, in a transformation model, a fragile and conflictual situation must be managed effectively to achieve business success. With the power to assist us in reshaping our world, our industries, and the fabric of how we live, Circular Science Cradle to Cradle® is a transformative approach. It's about reimagining our resources, designing products, remaking waste into non-toxic nutrients, and reinventing our economic models (see table Innovations 1986 – 2024 below).

It is evident within the swirling currents of change in industries that the pursuit of sustainability is an ongoing journey of transformation, not a fixed destination. Our role has evolved from innovating sustainable practices within a single textile company to advocating systemic changes across industrial sectors and geographical borders, highlighting how collective action and perseverance can reshape entrenched systems. Our steps accumulate into waves of change, but the quest for improvement persists.

The shift towards Circular Science Cradle to Cradle® is more than an environmental initiative— it's a comprehensive reimagination of businesses' operations. These pioneering companies are proving that profitability and sustainability can go hand-in-hand, creating a synergy that benefits their bottom line and the planet and its people.

Circular Science Cradle to Cradle® thus requires a seismic shift in mindset and philosophy. The corporate world is witnessing a groundbreaking movement of numerous companies not just joining the Circular Science Cradle to Cradle® initiative, but also becoming front runners as they trailblaze a path toward a sustainable and regenerative future.

Propelling Progress through Positive Chemistry

Throughout my multi-decade career, I have aimed to disseminate sustainability principles across industries and geographies. This has entailed persistence, collaboration, and envisioning possibilities beyond incremental improvements.

While advocating circular economy according to Cradle to Cradle® principles across industries, we have also collaborated with various partners to develop and implement positive chemistry solutions. These innovations in safer and sustainable chemicals have been instrumental in advancing circular products and processes.

Some pivotal chemistry pioneers have invented processes and products that should be highlighted in case studies so we can learn, anticipate, identify risks, develop business models, and translate them into tangible and successful actions.

Such case studies collectively illustrate that safer production is possible with foresight. They have advanced health, safety, and circularity across chemical value chains, especially in textiles, paper, plastics, wood, metal, electronics, and packaging. As this journey continues, boundless possibilities emerge for reimagining chemistry as a regenerative force.

While our collaborative efforts to develop positive chemistry solutions have been crucial, so too have been our policy advocacy, partnerships, and education for enabling the larger transition toward a sustainable, circular future across industries. But this remains an ongoing quest, not an endpoint. More innovation undoubtedly lies ahead as we align technology with Nature's wisdom.

The Future through Circular Science Cradle to Cradle®

Imagine a world where nothing goes to waste, where every product is designed non-toxic with its subsequent use in mind: a world where buildings generate positive renewable energy, clothes are integrated into endless closed-loop cycles, and electronics and all components and materials are designed for disassembly and reuse. It's a future of abundance, not scarcity, of regeneration, not depletion. Circular Science Cradle to Cradle Certified® makes this future possible.

Cradle to Cradle® and Circular Science

The Cradle to Cradle® design philosophy, central to Circular Science Cradle to Cradle®, focuses on creating products with a positive ecological impact that are safe for both human beings and the environment. Products certified under Cradle to Cradle Certified® are evaluated across five sustainability categories: material health, product circularity, clean air and climate protection, water and soil stewardship, and social fairness. This approach promotes circular life cycles for products and integrates innovative design with material health and reuse strategies, fostering a sustainable circular economy. Circular Science Cradle to Cradle® goes beyond mere recycling, representing a fundamental shift in our interaction with the world, harmonizing environmental, economic, and societal aspects. It advocates for a future where circularity is standard, guiding us to sustainably reimagine and reshape our world.[7]

---

[7] "Trademark Use Guidelines", Cradle to Cradle Products Innovation Institute, Accessed June 3, 2024, https://cdn.c2ccertified.org/resources/certification/policy/POL_Trademark_Use_Guidelines_20200930.pdf

The Future: Contours of a Circular Economy

The world in 2024 looks vastly different from the 1990s when we launched the first redesigned and upcycled textiles. Sustainable manufacturing is now a multi-billion-euro opportunity, driven by eco-conscious companies and consumers recognizing its value. However, linear production still dominates, indicating immense room for growth.

I envision a future where circularity is the norm across all industries. But change of this scale requires unified action from business leaders, policymakers, scientists, certification schemes, and citizens alike.

The transition will leverage emerging technologies like digital trackers, artificial intelligence, advanced recycling, bio-based and biodegradable materials, and green hydrogen. It also rests on localizing supply chains, educating consumers, and supporting small enterprises with financing and expertise.

Standing at this pivotal juncture, I am convinced that an equitable circular economy is the most viable pathway to balance industrial development with ecological regeneration. Its principles provide guideposts to create an economy where human progress resonates with the natural world's rhythms.

Innovators at the Forefront

What was once considered superior performance in a linear approach is often now perceived as "toxic" in a circular context. Innovative chemical mavericks are the alchemists of the modern age, who succeeded in transforming the process into "safe and circular chemicals." By innovating in chemical science, these researchers are creating sustainable and superior products in quality and functionality.

From Rebel to Radical Innovator

FROM REBEL TO RADICAL INNOVATOR:
Leading the Transformation through Circularity

INNOVATIONS 1986 - 2024

Legend:
Years: 0-9
Years: 10-19
Years: 20-30
still under way

#			Timeframe	Years	degree of fulfillment %			
	Chemical Industry Positive Lists	BN			25	50	75	100
1	communicated positive list chemicals	BN	1993 - 2010	17				
2	communicated positive list dyes	BN	1993 - 2016	23				
3	reactive dyes for cellulosic fibers	BN	1993 - 2013	20				
4	optical brighteners	BN	2004 - 2016	12				
5	printing inks textile industry	BN	2004 - 2018	14				
6	printing inks packaging	BN	2004 - 2016	12				
7	masterbatches safe for biological cycles	BN	2011 - 2015	4				
8	adhesives for wood	BN	2011 - 2015	4				
	Rawmaterials / Materials				25	50	75	100
9	biodragradable elastomer	BN	2002 - 2013	11				
10	biodegradable polymer	BN	1997 - 2013	16				
11	PET depolymerisation	TN	2004 -					
12	100% postconsumer plastics recycling „gold Level"	TN	2016 -					
13	Membranes for outdoor fashion	BN	1999 - 2016	17				
14	3 D biodegradable monofilament	BN	2015 - 2016	2				
	Products				25	50	75	100
15	Climatex	BN	1986 - 1987	2				
16	Climatex Lifecycle (safe for biological cycles)	BN	1993 - 1994	2				
17	Climatex Lifeguard FR (flame retardant)	BN	1994 - 1997	4				
18	Returnity PET CS	TN	2007 - 2008	2				
19	Wooden Parquet Flooring		2011 - 2015	4				
20	Toiletpaper	BN	2007 - 2010	3				
21	Offset printing paper	BN	2007 - 2018	11				
22	Flexible packaging	BN	2015 - 2018	3				
23	Packaging „gold"	TN	2011 - 2018	7				
24	Outdoor fashion	TN	1999 - 2019	20				
25	Workwear	TN	2011 - 2015	4				
26	Underwear	BN	2006 - 2010	4				
27	T-Shirt	BN	2004 - 2006	2				
28	Socks	BN	1989 - 2019	30				
29	Electronic goods	TN	2011 - 2024	13				
30	Beverages packaging Paper, Aluminium, Polyethylene alternative	BN	2015 -					
31	Cleaning detergents „gold"	BN	2011 - 2013	2				
	System Integration				25	50	75	100
32	Industrial composting commercially sound (after that approval authorities)	BN	1993 - 2018	25				
33	Circular economy	BN/TN	2007 - 2015	9				
34	Developing countries	BN	2005 - 2012	8				
35	Unido organization	BN/TN	2005 - 2016	12				
36	C2CPII	BN/TN	2006 - 2010	5				
37	Consortium	BN/TN	2007 - 2014	8				
38	Industry in entire country / region C2C (EU Commission)	BN/TN	2007 - 2017	10				
39	Entire production of a company Cradle to Cradle Certified	BN	2011 - 2015	5				
40	Business model for electronic goods	TN	2011 - 2012	2				

Figure 1: Summary of forty case studies: Transformation and innovation take more time than ROI, which by management rule is <1.5 years. In 38 years of business career, only two cases remain open.

Defining Circular Science

Positive Chemistry Case Study:

Positive List Chemicals (1993-2010)

Confrontation- Transformation

"Auxiliaries" are process chemicals that remain in the water during the manufacturing processes of washing or dyeing. Textile chemicals, by contrast, stay on the product to enable finishing properties, such as softeners and flame-retardants. The circularity, Cradle to Cradle® approach requires the properties of all chemical ingredients to be safe for biological and aquatic environments. For example, one of the challenges involves preservatives in the chemical formulation, which facilitate warehousing and transportation. Preservative properties, however, make the circularity requirement for biodegradability harder to achieve.

Thus, since cradle-to-grave linear thinking prioritizes pure performance, functional, quality, and financial goals, the environmental metrics in circularity and Cradle to Cradle® exist in opposition to conventional standards.

Commitment - Shaping Horizons

The search for alternatives that balance performance, and environmental safety is neither trivial nor could it be sidelined. The demand from the chemical industry was raised in 1993, but a natural system change solution was only offered seventeen years later, in 2010.

The story of Tanatex Chemicals is not just about chemicals. It's about reimagining an industry. Originating as a spin-off from the Bayer Group, evolving through Lanxess, and becoming independent in 2007 as Tanatex Chemicals B.V., Tanatex has a rich heritage that traces back to the invention of polyester in the 1950s.

In 2010, Tanatex Chemicals initiated an R&D and Marketing Workshop, Cradle to Cradle® Design Innovations, with Cradle to Cradle® scientists and experts to enable out-of-the-box thinking to confront the technical challenges. At the end of the workshop, they decided that their R&D approach would be driven by the integration of the circular science Cradle to Cradle® approach. A few months later, the first products with safe and circular standards were created, and these are widely used today within the textile industry. Competitors were suddenly more open to this approach and were able to extend their product portfolio after that. Today, a wide range of textile chemicals and textile auxiliaries are available in the global market.

Tanatex Chemicals' journey towards circular science Cradle to Cradle® shows the power of visionary thinking aligned with a commitment to sustainable development. It's a narrative that not only inspires but also demonstrates the practical approaches industries can adopt to meet the demands of a changing world, where environmental stewardship and economic success are not mutually exclusive but mutually reinforcing.

Positive Chemistry Case Study:

Positive List Color Dyes (1993-2016)

Confrontation-Transformation

Historically burdened with a significant environmental footprint, the textile industry faced escalating pressures to embrace sustainable practices. The traditional dyeing processes, heavily reliant on toxic pigments and extensive water use, posed substantial ecological challenges. Ancient dyeing methods used natural sources but was resource-intensive and costly. The challenge lay in identifying environmentally safe color dyes without compromising quality, which required deep scientific knowledge and extensive testing.

Commitment - Shaping Horizons

DyStar's origins date back to 1 July 1995, as a joint venture between Hoechst AG and Bayer Textile Dyes. Its capabilities expanded by incorporating BASF's textile dyes business in 2000. A crucial moment occurred in

2010 with its acquisition by Zhejiang Longsheng Group and Kiri Industries Limited (KIL), marking a new era of commitment to sustainable practices.[8]

DyStar's evolution from traditional dye producers to frontrunners in Circular Science Cradle to Cradle® was a superb example of innovation and visionary leadership. Their "Positive List Dyes" concept introduced dyes that were safe for society and the environment, transcending mere chemical replacement to embody the entire lifecycle of textile dyes. DyStar's efforts resulted in dyes that adhered to strict environmental standards, reducing the ecological footprint of textile production. This shift to sustainable practices, in line with circular economy principles, emphasized safety throughout the products' lifecycle, minimizing waste and environmental impact.

The results of DyStar's commitment were significant: a reduction in environmental impact, leadership in sustainable textile dye practices, and demonstration of the compatibility of environmental sustainability with economic viability. DyStar's journey was a testament to the transformative power of sustainable innovation in the textile industry.

Positive Chemistry Case Study:

Optical Brighteners (2004-2016)

Confrontation-Transformation

A hallmark of modern consumer markets is the widespread use of bright white color in fabrics, paper, plastics, and even personal care for hair and teeth. However, this trend raises concerns about effects on human health and the environment. In the realm of paper pulp bleaching, for instance, processes like Elemental Chlorine-Free (ECF) and Totally Chlorine-Free (TCF) each have environmental impacts. Optical brighteners that enhance the whiteness of cellulosic fibers, a fundamental requirement in the textile industry, pose especially complex

---

[8] "DyStar Group", Environmental Expert, Accessed June 4, 2024, https://www.environmental-expert.com/companies/dystar-group-139491

challenges. Traditional brightening methods often involve harmful chemicals, necessitating a reevaluation for safer alternatives, especially in the bleaching of polyester fibers, which are still not solved today.

Commitment - Shaping Horizons

Archroma's journey began in 1886 with the founding of Kern & Sandoz in Basel, a pioneer in textile dyes. The company's evolution continued through the formation of Clariant in 1995 as a spin-off from Sandoz and expanded with acquisitions, including Hoechst's specialty chemicals and BASF's textile chemicals business. In 2013, Archroma was established, marking a new chapter in its commitment to sustainable chemical solutions.

In 2004, Archroma took on the challenge of redefining optical brighteners for cellulosic materials. They aimed to align with the Cradle to Cradle® principles, aiming for C2C Certified Material Health Certificate™ Gold level compliance. This involved developing brighteners that were not only effective but also safe for both human beings and the environment. Archroma's transition from conventional chemical processing to sustainable optical brightening is a remarkable story of innovation, which drew on the company's rich heritage and collaborative efforts.

Positive Chemistry Case Study:

Color Masterbatches for Plastics (2011-2015)

Confrontation - Transformation

In the landscape of industrial evolution, Clariant and Avient represent a hallmark of innovation and sustainability. Originating from Sandoz's legacy in 1995, Clariant began transforming the chemical industry, specializing in care chemicals, catalysts, adsorbents, additives, and masterbatches.

The prevalent issue of "bad plastics," characterized by floating bottles and invisible microfibers contaminating our environment, has gained widespread public recognition. Coloring plastics, which require pigments as complex as those used in textiles, present unique challenges.

Though commonplace, the widespread use of color in plastics has raised concerns about the environmental and health impacts of traditional colorants. The challenge was dual: to develop color masterbatches that were vibrant, effective, and safe for society and the environment. This task extended beyond technical hurdles and became a matter of ethical duty. The 2019 acquisition of Clariant's color masterbatch business by Avient Corporation marked a crucial union of visions for a sustainable future, transcending a mere business transaction.

Commitment - Shaping Horizons

Clariant / Avient embarked on a transformative journey in the color masterbatch sector, aiming to create visually appealing and environmentally responsible products. Extensive research allowed them to use non-toxic ingredients, marking a departure from traditional practices. Their transformation from conventional chemical companies to circular science leaders in this sector showed how businesses can drive industry progress, harmoniously blending economics with environmental stewardship.

Innovation in Circular Science Cradle to Cradle®

Positive Chemistry Case Study:

Reactive Dyes for Cellulosic Dyeing (1993-2013)

Confrontation - Transformation

Reactive dyes, used by approximately 98 percent of global textile dyehouses for cellulosic fibers and fabrics, posed significant environmental challenges from a Cradle to Cradle® perspective. These reactive dyeing pigments, with their chemical anchors, were harmful when introduced into biological cycles. Traditional dyeing processes for cellulosic fibers were marked by high water and energy usage, considerable waste, and reliance on detrimental chemicals. Since the pigments can disrupt nature's biological cycles, such practices increasingly clashed with rising sustainability standards and changing consumer expectations.

DyStar was established in 1990 through the amalgamation of the textile dye businesses of Hoechst AG and Bayer. Recognizing the need for more sustainable dyeing solutions in the early 1990s, DyStar embarked on a journey to develop reactive dyes that were both effective and environmentally responsible.

With few alternatives available, innovation became imperative. Despite market pressure to accept reactive dyes, the Cradle to Cradle® scientific review maintained its stringent criteria, refusing to compromise. An alternative, VAT dyes (water insoluble), required unique equipment and processes, limiting their adoption to only about 2 percent of dyehouses.

Commitment - Shaping Horizons

While achieving significant progress, DyStar faced challenges like market resistance, the necessity for customer education, and the complexities of integrating new dyes into existing manufacturing systems. Through persistent efforts and stakeholder collaboration, these hurdles were eventually overcome.

Their dedication to creating environmentally safe products has not only advanced the textile industry but also set benchmarks for other sectors. DyStar's legacy continues to inspire future sustainable manufacturing innovations and responsible consumption, with multiple companies now offering reactive color dyes globally.

Positive Chemistry 2 Case Studies:

Printing Inks for Textiles (2004-2018)
Printing Inks for Packaging (2004-2016)

Confrontation - Transformation

Printing inks in textiles and packaging, despite their aesthetic contributions, are laden with chemicals that can damage the environment and human health. This presented a paradox within the vibrant textile and packaging industry. The diversity and complexity in printing technologies and materials required inks with specific properties, challenging Textilcolor and Siegwerk to balance technical demands with international safety standards.

Recognizing the critical importance of material health and circularity, Textilcolor committed to revolutionizing textile printing inks. The company initiated a drive to formulate inks and coatings that were environmentally safe and met societal needs. This effort required a thorough reevaluation of the many components of the ink manufacturing process, so harmful chemicals could be eliminated while maintaining quality and performance.

Shaping Horizons

Textilcolor exemplifies how industry challenges can catalyze innovation and sustainable growth, with the company moving from traditional chemical production to cutting-edge developments for textile printing while adhering to circular science principles.

Textilcolor envisions innovations in textile chemicals, dyes, and print framing applications engineering, product development, and product safety with a team in over forty countries worldwide.

In the heart of the packaging printing industry, a story of transformation and commitment unfolded with Siegwerk, a family-owned business in its sixth generation with nearly two centuries of printing expertise.

Siegwerk's story showcases the blend of heritage and innovation, evolving from classic ink manufacturing to a frontrunner in circular science for packaging inks. This transition underlines the power of established expertise in driving change, illustrating the balance between environmental stewardship and financial viability, while contributing to a sustainable future.

Toolbox Cradle to Cradle™ Methodology

Positive Chemistry Case Study:

Adhesives, Make it happen with an integrated system approach (2011-2015)

Confrontation - Transformation

Crucial in various manufacturing sectors, the adhesive industry has historically faced environmental challenges due to harmful substances in traditional adhesives. Alfa Klebstoffe recognized the need for sustainable adhesives that are robust, reliable, and environmentally safe. Established in 1972, ALFA Klebstoffe AG, from the beginning, developed environmentally friendly adhesives and consistently pursued innovation. To this day, ALFA Klebstoffe's vision dominates the further development of adhesive products, which are exported to over 90 countries. As a result, ALFA Klebstoffe AG is one of the market leaders in the field of environmentally friendly, water-based adhesives. The ALFAPURA range, created using the Toolbox Cradle to Cradle™ approach, exemplifies practical, environmentally friendly adhesives, supporting the principles of a circular economy. These adhesives are designed to be safe for biological systems, ensuring no harm to the biosphere.

Product design today often involves combining different materials, each needing a unique adhesive formula. This trend complicates product integration into certification schemes. The Toolbox Cradle to Cradle™ was developed to address this problem, incorporating the supply chain of materials, chemicals, colors, inks, and components. This toolbox applies a worst-case scenario approach, which considers the maximum potential weight used in a product. Once new suppliers or ingredients aren't added, products derived from the Toolbox Cradle to Cradle™ are automatically eligible for Cradle to Cradle Certified® or C2C Certified Material Health Certificate™, which is applicable to products with many inputs, such as paper products, textiles, wood products and more.

Alfa Klebstoffe's transition to safe and circular adhesives has simplified sustainability for manufacturers, highlighting the practicality of environmental responsibility in the industry. Somewhat surprisingly, the restrictions imposed by the Toolbox Cradle to Cradle™ methodology didn't limit inventiveness but instead opened new avenues for innovation, eliminating concerns over harmful chemicals.

Commitment - Shaping Horizons

Alfa Klebstoffe adopted a systematic and innovative approach to addressing industry challenges – the Toolbox Cradle to Cradle™ Methodology. This strategy involved more than developing new adhesives; it was a complete reevaluation of the bonding process.

A notable aspect of the ALFAPURA adhesives was their seamless compatibility with existing bonding processes, allowing manufacturers to transition to safer, circular options without needing to revamp their current systems.

Positive Chemistry Case Study:

Fragrances are a highly confidential industry; how do you get the yes? (2019 - 2023)

Confrontation - Transformation

In the fragrance industry, where confidentiality and exclusivity are paramount, LUZI faced the challenge of crafting captivating scents in an environmentally responsible way. The Toolbox Cradle to Cradle™ Methodology provided a solution that upheld the highest levels of confidentiality while fostering creativity and innovation.

LUZI stands as a paragon of innovation and sustainability in the enchanting world of fragrances. Founded in 1926, this family-owned Swiss company carved a niche in developing unique scents for fine fragrances, personal care, and home and air care products. Between 2019 and 2023, LUZI embarked on a transformative journey to create captivating aromas and redefine the fragrance industry's approach to sustainability.

LUZI concentrated on creating fragrances that were not only unique and appealing but also environmentally friendly. This required a careful selection of raw materials and eco-friendly processes. Navigating the industry's need for confidentiality, LUZI successfully incorporated sustainable practices without sacrificing the secrecy vital to their clients. This balance between transparency and discretion was key to their success.

Their innovative use of the Toolbox Cradle to Cradle™ Methodology went beyond mere fragrance formulation—it integrated material health into the basics of scent creation. LUZI fragrances are C2C Certified Material Health Certificate™ certified Gold.

Commitment - Shaping Horizons

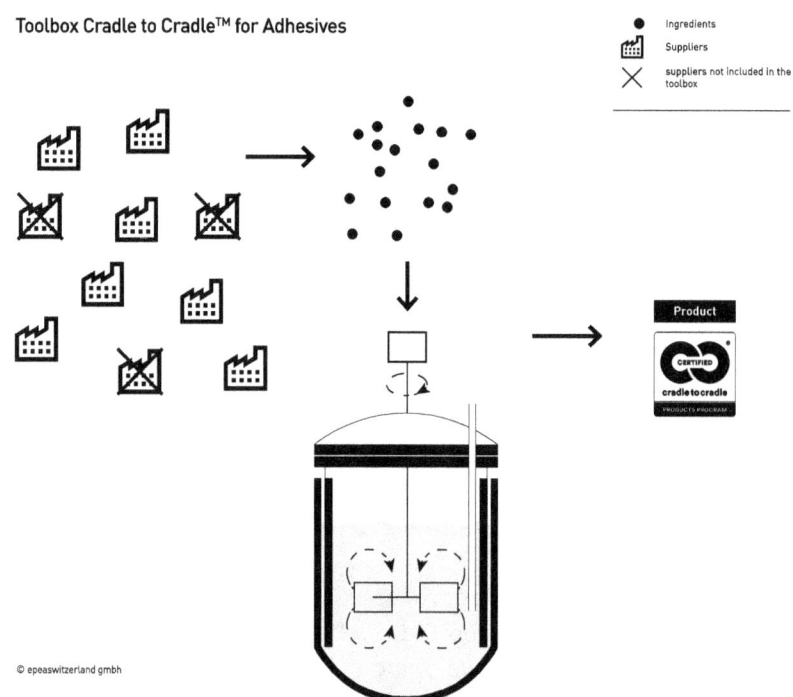

Figure 2: Toolbox Cradle to Cradle™. With this tool a wider range of products with more flexibility in manufacturing can be generated.

A Safe and Circular Cradle to Cradle® Future Defined by Innovation and Sustainability

The companies we have spotlighted do not merely represent a niche trend; rather, we're witnessing a global movement. These trailblazers are setting new standards, showing that the path to success in the 21st century is paved with innovation, responsibility, and a deep respect for our planet.

Circular Science Cradle to Cradle® is an invitation to rethink, redesign, and rebuild. It's a call to action for businesses to be, not just profit-driven entities, but forces for good, champions of change, and stewards of the Earth.

We extend our heartfelt admiration and support to the companies joining this revolution. Their journey is not just inspiring—it's essential. They build a future where business and nature thrive in harmony, where every action is thoughtful and every strategy sustainable.

Let's redefine what it means to be successful in business and create a legacy that future generations will look back on with pride and gratitude.

An Ongoing Journey of Impact

My sustainability journey continues to unfold through collaborations with partners who share a safe and circular science vision. The challenges have multiplied, from reinventing a single textile mill to transforming complex global supply webs. But so have the opportunities.

The work I began with Climatex Fabrics has cascaded into new industries through the creativity of entrepreneurs, designers, engineers, and young people who have amplified these ideas. Witnessing this momentum is profoundly gratifying.

Of course, there are frustrations when progress seems to stall due to incrementalism. But I remind myself that enduring change starts small before reaching tipping points. Our task is to persevere with purpose.

My mission now is to steward these ripples of transformation while empowering the next generation of sustainability trailblazers. Ultimately, any lasting legacy will be measured by how many carry this mission forward.

Reflecting on the past three decades, it is astonishing to feel the groundswell of support compared to the early skepticism. But there is still so much to accomplish. I hope that in another thirty years, the ideas we pioneered have become inevitable, shaping a world we proudly hand down to posterity.

CHAPTER 4

Transforming Materials – Safe and Circular Materials - Case Studies

Transformation from Negative to Positive Definition Case Study

Defining Elastomers (2002-2013) & Polymers (1997–2015) Safe for Biological Cycles

Confrontation – Transformation

Today, 60 percent of textiles are made from blends of cotton with polyester (PET) or cotton with elastomers, primarily polyurethane-based (PU-based) fibers and yarns. This mix of biological cotton and technical PET and PU fibers presents a significant Cradle to Cradle® design challenge in circularity, as these materials cannot be easily separated without degradation in quality or downcycling. Despite various technological initiatives to address this, they are often seen as end-of-pipe solutions, attempting to fix a fundamentally flawed system. The scale of this issue is vast: in 2022, global cotton production reached 27 million tons, while polyester fiber production hit 63.3 million metric tons, a substantial increase from the 3.37 million metric tons produced in 1975.

The environmental impact of fiber production is also alarming. Global herbicide consumption reached 1.7 million metric tons in 2021, with pesticides used extensively in cotton farming. Cotton, occupying 2.5 percent of agricultural land, is one of the most pesticide-intensive crops, consuming 16 percent of all insecticides and 7 percent of all herbicides globally, more than any other crop. Polyester (PET) also poses environmental hazards as the material does not biodegrade, leading to issues like microfiber and microplastic pollution. Polyester production also requires antimony, which is a heavy metal and carcinogenic substance that poses significant risks to human health, society, and the environment.

These factors underscore an urgent need for effective solutions and a rapid global transformation in textile production practices. To give a simple example, in the manufacture of socks, high-performance features include stretch functionality and reinforcements for toe tip and heel. But using 100 percent natural fibers would not allow such applications. The same is true in the manufacture of workwear garments with robust properties. The solution to this dilemma is to innovate synthetic fibers that are safe for biological cycles, which can biodegrade even in the worst-case scenarios. This new generation of materials would need to have excellent performance at both the manufacturing stage and in usage. Is this an exaggerated, impossible, unachievable dream? Transforming the largest, most polluting global industry by taking the entire textile and clothing industry by its horns and giving it a firm shake and a wake-up call? Everyone ridiculed, negated, and rejected this approach as crazy, impossible, unfit for purpose, and over the top. But that's precisely why a rebel rebels. He needs that confrontation to prevail.

Without proof that innovative materials meet performance metrics and are safe for biological cycles, any further path to integrating and transforming the global textile and clothing industry will be impossible.

How do you find allies in the chemical industry who want to take on this risk, go for the adventure, and who can also provide the financial resources despite the resistance? It seems like a hopeless venture doomed to failure.

Defining Elastomers (2002-2013) Safe for Biological Cycles

The Fein Elast Group specializes in manufacturing elastic core yarns. During 2002 to 2007, in my tenure on the company's supervisory board, I saw an opportunity to propose the development of elastomers safe for biological cycles to the company's C-level executives. Initially, the idea was met with skepticism and some confusion. It required persistent efforts to secure approval for the project, which was anticipated to be a long-term initiative, potentially spanning years or even decades. In retrospect, persuading the management that this development could be pioneering was a significant achievement.

Founded in 1966, Fein Elast Group has established itself as a premier international elastic and non-elastic combination yarn manufacturer. With its 250 employees across Austria, Switzerland, Germany, and Estonia, the company expanded its product range from fashion to medical stockings, home textiles, automotive, and aerospace. The group's diverse yarn offerings included single and double-covered yarns, air-interlaced, Hamel Elasto Twist, Corespun, Sirospun, twisted, Knit-de-Knit, and metallic yarns. These products have found a market in nearly 50 countries worldwide.

Thanks to the introduction by Fein Elast Group, we established contact with a significant elastomer yarn supplier in 2002. The development journey, lasting an arduous thirteen years, was filled with setbacks, disappointments, and continual new challenges. The project's complexity stemmed from various factors, including using alternative chemistry to the polyurethane process, maintaining technical performance throughout production and usage, and establishing scientific proof of safety for biological cycles. Validating scientific hypotheses within a biological context required extensive, time-consuming, and costly testing in specialized laboratories. These continuous challenges tested everyone's patience to the limits.

Defining Polymers (1997–2015) Safe for Biological Cycles

In 2011, Lauffenmuehle, a comprehensive textile manufacturer known for high-quality performance workwear fabrics, conducted a Cradle to Cradle® Design Workshop, applying the Reference Model Cradle to Cradle™. Initially, it seemed all avenues were dead ends, with no apparent solution for creating products that combined functionality, high performance, and safety for biological circularity. Nonetheless,

Lauffenmuehle's CEO approved the project's budget, recognizing the potential value and opportunities of such an innovation. This decision came when the company was striving to stay afloat in a highly competitive global market. The project team, driven by enthusiasm, professionalism, and deep respect for future challenges, began integrating innovative practices into Lauffenmuehle's comprehensive textile operations, including spinning, weaving, dyeing and finishing.

Lauffenmuehle, located in the south of Germany's Black Forest near the Swiss border, was one of Germany's largest vertically integrated textile companies in the 20th century, encompassing spinning, weaving, dyeing and finishing. Since its founding in 1835, Lauffenmuehle faced numerous challenges, including five insolvency proceedings since 1993, reflecting the broader decline of Germany's textile industry. The company maintained operations until 2019.

In 2013, a breakthrough occurred when a plastic polymer, initially designed for a non-textile product group, was identified. Extensive feasibility studies and trials were conducted to explore the possibility of using this material for producing fibers and yarns. Scientists evaluated the polymer's ingredients and chemistry against Cradle to Cradle® standards for biological system safety. Although the fiber and yarn outcomes were promising, the material still posed significant processing challenges for the textile industry. Moreover, a setback occurred when the material was found to contain a carcinogenic substance, which necessitated a phase-out and redesign of the polymer.

Convincing the polymer supplier to modify the formula to align with the "safe for biological cycles" concept involved complex, contentious discussions about scalability, credibility, and professionalism. Eventually, an agreement was reached, including aspects of exclusivity. This marked a pivotal moment, transforming what seemed an impossibility into a tangible reality and laying the foundation for further development.

The advancement, however, required numerous innovations in process chemicals, colors, and finishing chemicals, as well as technological advancements in spinning, weaving, and finishing. Developing trimmings suitable for apparel applications was also part of this comprehensive innovation process.

Commitment – Shaping Horizons

In the following section, I highlight three innovative products from different companies, each showcasing advancements in sustainable textile materials.

In July 2017, Asahi Kasei Spandex Europe GmbH, the German subsidiary of the Japanese company, made a significant announcement about the development of ROICA™ V550. This spandex product, environmentally less harmful and safe for biological cycles, achieved the C2C Certified Material Health Certificate™ Gold Level for yarns produced in Germany.

Asahi Kasei Corporation, a Japanese multinational chemicals and materials science company, was founded in 1922. It initially focused on ammonia and nitric acid production. Significant milestones in its history include a joint petrochemical venture with Dow Chemical in 1952 and diversification into various sectors like construction materials, electronics, and pharmaceuticals from the 1960s to the 1990s. As of March 2023, Asahi Kasei employed 48,897 people and operated fifty-four manufacturing facilities globally.

By 2022, production in Germany ceased, transferring knowledge and operations to Asahi Kasei in Japan, which also received the C2C Certified Material Health Certificate™ Gold Level. Asahi Kasei Group's longstanding commitment is to contribute to societal development and anticipate emerging needs, embodying their ethos of "Creating for Tomorrow."

Following the closure of Lauffenmuehle in 2019, its former CEO founded Inogema GmbH. Inogema focuses on transforming today's products into tomorrow's resources.

Inogema GmbH presented Vinatur® Workwear fabric and accessories designed for biological safety and offering climate control functions. These products were ideal for industrial laundering, demonstrating excellent care, light fastness, and resistance to abrasion and pilling. Vinatur® Workwear was not only suitable for workwear performing after more than fifty cleaning cycles, but also for other high-performance textile products. Products are certified at the gold level by Cradle to Cradle Certified.

In 2022, Inogema sold its raw material yarn business to OceanSafe AG. OceanSafe started in 2019 in Switzerland and Germany and is the brainchild of Manuel Schweizer, who brought over a decade of R&D experience in circular interior textiles and three decades in the textile industry. OceanSafe's technology is designed to address the environmental challenges of the textile industry, helping brands and retailers develop circular textiles. Embracing the Cradle to Cradle® product design philosophy and achieving the Cradle to Cradle Certified® Gold Level, OceanSafe now licenses its technology across the textile value chain, pioneering a sustainable business model in the industry.

In 2022, OceanSafe introduced coNea, a synthetic textile material designed as an alternative to conventional cotton. naNea, a new polymer that could be derived from biobased feedstock through fermentation, addressed the environmental and health concerns associated with traditional polyester. It was designed for biodegradation without additives and was free from hazardous substances.

As this innovative material is still in the early development stages, its future in achieving global acceptance, penetration, and scalable success remains to be seen.

Fighting against Goliath; Case Studies:

PET Depolymerization, phasing out Antimony (2004–open) & Membranes for Outdoor, Lifestyle Fashion, phasing out PTFE (1999–2016)

Confrontation – Transformation

Polyethylene terephthalate (PET) plastic, first synthesized in the 1940s by DuPont chemists, was initially developed as an ideal material for textile fibers. It wasn't until 1973 that the first PET bottle was patented, designed with a linear approach that didn't consider using materials across multiple lifecycles or their safety for humans and the environment. In 1977, PET bottle recycling began, eventually becoming widespread, leading to its extensive use in textile products and beverage packaging. However, significant issues with PET have since emerged. Polyester does not biodegrade, leading to the proliferation of microfibers and microplastics. Moreover, antimony, a heavy metal and carcinogenic substance used as a catalyst in PET production, poses additional environmental and health risks.

Businesses and industries comply with legislation, so they do not need to fear any legal penalties. By law, companies are required to list all toxic ingredients in a product > 1000 ppm (parts per million) (0.1%). Antimony is present between 200 and 300 ppm (0.02-0.03%) and the legal limit for antimony migration is 40 mg/liter–effectively, it is 6-12 mg/liter. Since the industry meets the required margin, the authorities have declared virgin PET and the packaging of recycled PET beverage bottles qualify for food grade, giving businesses the green light to use them without restrictions for drinking water bottles. Within a Cradle to Cradle Certified® certification, however, only a bronze level would be achievable because of the presence of carcinogenic, mutagenic, and reprotoxic (CMR) substances.

In the transformation process, we ask: can we repair the existing PET material streams, making them safe for biological systems and replacing the cancerogenic antimony with an environmentally friendly catalyst? Chemical depolymerization would be required to exchange elements that need to be phased out. The investment for a chemical upcycling plant would be an estimated US$200 million. But no investor could be found because the virgin PET material is approximately US$1/kg, making it impossible–to break even in the business model for depolymerization (chemical recycling). However, by adopting a repair approach, recycled materials could surpass virgin materials in quality, achieving true upcycling.

In 1946, DuPont introduced polytetrafluoroethylene (PTFE) Teflon to the world, significantly impacting the lives of millions while simultaneously introducing pollutants into their bodies. Today, the family of compounds, including Teflon, known as per- and poly-fluoroalkyl substances (PFAS), is present in various products, from pots and pans to carpets and textile membranes. Alarmingly, PFAS chemicals have been found in the blood of people globally, including in 99 percent of Americans. These chemicals, which pollute water and do not break down, persist in the environment and human bodies for decades, earning them the moniker "forever chemicals." (see Chapter 1).

Membranes in textiles and shoes block liquid water but allow water vapor to pass through. The products are claimed to remain dry. It is designed to be lightweight, waterproof fabric for all-weather use. It comprises expanded PTFE (ePTFE), a stretched-out form of the PFAS compound polytetrafluoroethylene (PTFE). The membranes are bonded with several layers of different materials, making them impossible to circularize. From a design perspective, these products are highly toxic and wasteful. Products containing PTFE or PFAS cannot be certified Cradle to Cradle Certified®.

A few global players split the market among them, and their power is phenomenal. This makes it extremely hard for others with better and alternative solutions to enter the market. Now, however, authorities are putting pressure on these global players to innovate environmental solutions. There might be a chance for newcomers and start-ups to get a piece of the pie.

Commitment – Shaping Horizons

We have sought global partners to champion this cause for over two decades, yet it remains largely unnoticed and unaddressed. We have demonstrated in numerous projects that PET production can be safe and circular for biological systems, as further detailed in other case studies in this book. However, it remains a mystery why significant players in the beverage industry and the vast textile and fashion sectors overlook these facts and fail to transition to healthier and more circular materials. Most consumers are unaware of these underlying issues, but increased awareness could lead to public demand for change and solutions.

In 2016, during one of our Cradle to Cradle® Textile and Fashion Projects, we discovered a practical solution for hydrophobic finishes, replacing PTFE and PFAS within our Network of Trust™, achieving a C2C Certified Material Health Certificate™ Gold Level. Interestingly, the chemical company that contributed to this breakthrough did not come from the textile sector but was in the cleaning and detergent industry. Werner & Mertz, with their Modular Water Repellent System, and the Grabher Group, a textile finishing company incorporating innovative technology, played critical roles in this project. The partners filed for a patent, which took five years to secure, underscoring the competitive nature of this market. This innovation was successfully integrated into a project for a major fashion brand. The performance results were satisfactory for regular consumer use, though not for extreme activities like scaling Mount Everest.

Large market-leading companies often create barriers, increasing market pressure and challenging market entry for newcomers. However, with new legislation on the horizon, there is hope for change. I remain optimistic about the potential for these advancements.

Background

We all know Goliaths; they are present throughout our daily lives.

An estimated > 500 billion PET packaging bottles are produced each year. But the most significant amount of PET is used in textiles and fashion production.

The membrane market, mainly PTFE and PFAS, is shared among a few global players and was estimated at over US$ 6 billion in 2022. In 2023, US and EU legislators announced that they are in the process of banning or putting PTFE and PFAS products on the watchlist—a terrible disaster for an unprepared industry.

Werner & Mertz is a family-owned business in the fifth generation. With famous brands, particularly trusted ones like Frosch and Green Care Professional, the corporation is recognized as an innovative market leader for cleaning products and laundry detergents.[9]

---

[9] "About us", Werner & Mertz, Accessed June 3, 2024, https://werner-mertz.de/en/about-us/

The Grabher Group is a leading manufacturer of high-tech products. As a manufacturer of technical textiles, the Grabher Group is always looking for new challenges. Their core competencies are in producing filter media and intelligent textiles.

Developing the Business Case; Case Studies:

Single Use Plastic for multiple uses (2016) & 100 percent Postconsumer Textiles, Chemical Depolymerization (2014–2018)

Confrontation – Transformation

The widespread use of single-use plastic products (SUPs) has triggered significant environmental and health problems. Single-use plastic products, wholly or partially made of plastic, are designed for single-use or short-term use before disposal. These plastics, often found in our oceans, comprise 70 percent of marine litter on European beaches, predominantly comprising the top ten single-use items and fishing gear. The environmental impact of such waste is a global concern, with single-use plastics more prone to ending up in the seas than reusable alternatives.

In response, legislation is stepping in to provide a regulatory framework. The EU, taking action against plastic pollution, implemented measures on 3 July 2021. These include banning single-use plastic items like plates, cutlery, straws, balloon sticks, cotton buds, containers made from expanded polystyrene, and all oxo-degradable plastic products in EU member states. The legislation also aims to reduce the usage of specific plastics where alternatives are available.

This legislative shift has compelled industries and brands to innovate and achieve compliance. A historic example of this evolution can be seen in dairy product packaging. Once made of waxed paper or glass, milk bottles transitioned to blow-molded plastic variants in the 1960s. High-density polyethylene (HDPE) and polyester are commonly used materials in this context, exemplifying the shift from traditional packaging to modern plastic solutions and highlighting the ongoing journey towards responsible and sustainable packaging. Collecting and recycling systems for dairy products are widely used. While they vary from country to country, plastics are recycled and used as a reliable source for other nonfood packaging applications.

In 2016, we embarked on an ambitious project to create a closed-loop system where dairy bottles are continuously recycled into new dairy bottles. Crafting this business model was complex. We established material health criteria aligned with the Cradle to Cradle Certified® gold-level standards for plastics, colors, printing inks, and additives. These criteria were adapted for a diverse manufacturing supply chain, accommodating different legislative requirements across multiple countries. A significant aspect of this project was to understand consumer behaviors and to ensure efficient packaging return, which is crucial for reintegrating raw materials into the supply chain and achieving true circularity. However, illustrating the financial feasibility of this concept proved challenging with traditional linear accounting methods, leading us to adopt the Circular Accounting methodology by epeaswitzerland™, as detailed in Chapter 12.

Recycling is predominantly mechanical, often resulting in more expensive and downcycled products from a Cradle to Cradle® perspective. Despite this, there's a growing consumer willingness to pay a premium for products containing recycled content. The complexity of global supply chains, where materials like plastics, metals, and wood contain varied and often toxic chemical compositions, poses a significant challenge. Purifying these materials necessitates chemical processes, such as plastic depolymerization, which strips away all additives, catalysts, and colorants. While costly and energy-intensive, this technology is essential for achieving material purity. Aquafil has managed to navigate these challenges, emerging as a global leader and a successful example of sustainable business practice.

ECONYL® exemplifies this new approach. Produced entirely from waste materials like old carpets, fishing nets, fabric scraps, and industrial plastic, ECONYL® can be infinitely recycled while maintaining high quality. Importantly, it offers a significantly lower environmental impact in CO_2 emissions than traditional nylon.

Commitment – Shaping Horizons

Proof of concept for the dairy packaging project allowed an accurate estimate of the volumes needed for a closed-loop system to break even financially. Manufacturing trials, which involved cleaning, grinding, reprocessing, and molding the recycled plastics, showed promising results. Implementation was the next step. However, after a thorough risk assessment by management, the project was discontinued. Factors like investment, economic viability, technology, collection infrastructure, consumer behavior, and evolving legislation all made the project unviable. Despite this, the pressure on businesses to adopt sustainable practices continues to be a significant incentive for innovation.

As of 2022, ECONYL® regenerated nylon, fashion and textile, thanks to its unique characteristics, has supported more than 2,500 brands in their journey towards sustainability. Aquafil Econyl® is a C2C Certified Material Health Certificate™ at the gold and silver levels. The materials are successfully integrated in epeaswitzerland's projects Wolford and Napapijri.

In 2022, Aquafil launched the first demo plant to produce bio-based nylon on a pre-industrial scale. The successful pilot phase proves that it is possible to break our fossil fuel-driven system and transform our value chain to be sustainable, regenerative, biobased, and circular.[10]

Background

The common thread linking the EU Packaging and Packaging Waste Directive, dairy products legislation, and the creation of ECONYL® nylon by Aquafil S.p.a. was their collective focus on environmental sustainability and responsible production practices.

The 1994 EU Packaging and Packaging Waste Directive, revised in 2018 and enacted in 2020, became a crucial component of the EU Green

---

[10] "Aquafil Sustainability Report 2022", Aquafil, Accessed June 4, 2024, https://www.aquafil.com/assets/uploads/ENG_RS_Aquafil_2022.pdf

Deal and Circular Economy Action Plan. This legislation aimed to reduce the environmental impact of packaging waste, promote sustainability in packaging materials, and encourage recycling.

Simultaneously, dairy products legislation ensured the safety and hygiene of food production, setting standards for microorganisms, chemicals, and food additives, and included requirements for product labeling. This legislation underscored the importance of consumer safety and environmental health in food production and packaging.

Lastly, the connection to Aquafil S.p.a.'s ECONYL® nylon exemplified the practical application of these principles in industry. As a leading manufacturer, Aquafil embraced the circular economy concept, especially with its flagship product, ECONYL® nylon. This innovative material, created from recycled waste, demonstrates how industries can contribute to sustainability by reducing waste and reusing materials, aligning with the broader objectives of the EU directives.

Together, these areas—packaging waste management, food safety and labeling, and sustainable industrial practices—showcase how legislation and industry innovation can work hand in hand.

Developing High Tech Materials; Case Studies:

Luxury Glass Crystals (2023-2024)

Confrontation – Transformation

Can the allure of luxury products coexist with sustainability? Is it possible for high-tech materials to be safe and circular in material health and product lifecycle?

Swarovski, a renowned Austrian brand, is at the forefront of marrying luxury with sustainability. Since 1895, founder Daniel Swarovski's mastery of crystal cutting has defined the company. His enduring passion for innovation and design made Swarovski the world's premier jewelry and accessory brand. Today, the company carries on the tradition of beautiful products of impeccable quality and craftsmanship that bring joy and celebrate individuality, as part of the LUXignite strategy. Swarovski crystal settings come to life through proprietary techniques, giving handcrafted work a highly precise characteristic. Their attention to detail ensures that every piece will always be made of the highest quality. This expertise is infused in every step of the creative process, from initial design sketches through assembly. Since its founding, Swarovski has been committed to philanthropy and sustainability, continuously investing to preserve the environment and positively impact global and local communities.

In a significant move in 2024, Swarovski introduced Swarovski ReCreated™ crystals, Swarovski's most sustainable crystals to date. Made using crystal breakage from its manufacturing processes that is remelted into new crystals, this innovation uses at least 40% less natural resources compared to standard Swarovski crystals.

Swarovski ReCreated™ crystals combine Swarovski's savoir-faire and their spirit for circular innovation. The company is working on embedding these new crystals across their collections, continuing to improve the sustainability of their products.

Commitment – Shaping Horizons

The next step of innovation is the ambition to transform design, production and consumption practices in order to preserve resources and reduce waste.

3D Printing Monofilaments (2015-2016) Safe for Biological Cycles

Confrontation – Transformation

Integrating new manufacturing technologies into the contemporary world poses its own set of challenges. 3D printing, for instance, has the potential to revolutionize the industry by decentralizing manufacturing processes. It could significantly diminish the need for extensive logistics and warehousing, while providing efficient means for tooling, repairing, prototyping, and creating new products. However, a critical question remains: What is the safety and circularity of materials used in 3D printing? Many materials currently available in the market raise concerns due to their toxicity and incompatibility with circular economy principles. This issue needs addressing to ensure that 3D printing technology can fully align with the goals of a safe and sustainable environment.

In 2015, a feasibility project was initiated with the Network of Trust™ of epeaswitzerland to manufacture a 3D extrusion printing monofilament using the innovative polymer vinatur® from Inogema and naNea from OceanSafe. Once this was available, the project was extended, adding applied science and an intelligent textiles network platform to manufacture 3D printed products that could be integrated into Cradle to Cradle® projects.

Commitment – Shaping Horizons

The inaugural product developed was a glider for home textile curtains, which achieved market success after its launch by the Swiss home furniture retailer Pfister. These curtains were distinguished by the Cradle to Cradle Certified® certification at the gold level. While plastic molding solutions for this glider would have been more economically viable, the commitment to using safe and circular materials presented manufacturing challenges that were successfully navigated.

The second project, to develop a zipper for fashion using 3D technology, originated from a dilemma: the novel properties of the raw materials made standard manufacturing technologies impossible to use. Basically, the raw materials are too unstable and soft to make a functional slider in the zipper. The challenges have not yet been solved.

CHAPTER 5

Revolutionizing Education - Embracing a New Ethos in Industrial Design and Economics

In this chapter, we turn the spotlight on education, focusing on changing mindsets. At the heart of this discourse is the concept of industrial design. Traditionally, this realm has been characterized by linear thinking. While creativity abounds in product creation, principles like non-toxicity, circular design, and material reutilization embody an emerging imperative. Industrial designers need to be champions of this new ethos, crafting sustainable products that can be repurposed without sacrificing quality.

This paradigm shift is a professional requirement and a fundamental aspect of designers' educational journey. Economics is the next frontier where linear thinking still reigns supreme. The call of the hour is to transition towards a circular mindset. Consider the traditional approach to accounting—it's linear. We propose a pivot to circular accounting as a tangible demonstration of circularity's economic viability, a concept often overlooked in the conventional model.

Then, there's sustainability, a buzzword in contemporary discourse. The standard reference point here is life cycle assessment (LCA), which, despite its merits, remains a linear model. LCA evaluates the environmental footprint of products throughout their lifespan. In contrast, Cradle to Cradle® is an innovative framework to preserve raw material quality within a circular economy. LCA operates because all products inevitably lead to pollution and waste. Cradle to Cradle® implies that products can be designed to positively impact people, the environment, and even the economy.

These diverse elements underscore why revolutionizing education is paramount. When examining Cradle to Cradle® in professional circles, it becomes evident that experts across various fields are adopting circular thinking. Many start with a linear approach, attempting to retrofit it into circularity, but often hit a wall due to persistent waste issues. The real breakthrough occurs when they completely eradicate the notion of waste.

The significance of education cannot be overstated, especially for professionals who possess deep expertise in their fields. The transition to circularity is a challenging feat given their deep-seated knowledge of traditional methodologies, and so needs a guiding hand. This is where "Cradle to Cradle for professionals™" comes into play, emphasizing the pivotal role of education in propagating new concepts.

A compelling insight emerged from university research featuring a cooking book project. The findings were eye-opening: younger children aged five to eleven effortlessly grasped product design and circularity concepts. Conversely, older children aged twelve and above found these concepts more challenging to assimilate, highlighting how education molds mindsets.

This methodology has also been effective with professionals, dating back twenty-five years with a group of architects. Through design workshops, they were able to conceive their own products. The feedback has been overwhelmingly positive, validating the impact of active learning and real-world application. Such experiences foster creativity and the practical application of circular principles.

Drawing from personal experience, I recall the challenge of being forced to write with my right hand despite being naturally left-handed. This personal rebellion against the status quo reflects the broader difficulty in shifting educational systems entrenched in linear thinking, especially at students' age. In my generation, using the left hand was stigmatized. This experience left an indelible mark on me, fueling my skepticism towards accepted norms and underscoring the importance of commencing educational transformation with younger generations, who are more malleable in their path choices. It's more feasible to mold young minds, as their beliefs and habits are not firmly established.

The challenge escalates with age as individuals become more deeply rooted in their existing mindsets. Recognizing the adaptability of children to new concepts and the need for professionals to gain a deeper understanding, we conceived the theme dealt with in Chapter 7, "Transforming Design." We introduced a Reference Model Cradle to Cradle™, comprising eighteen points (akin to a golf course) to guide professionals. This model has proven effective in orienting businesses and aiding their transition to circularity.

A case in point is the Froggy brand's cleaning pouch, which illustrates the impact of integrating Cradle to Cradle® methodology into product development and communication. This underscores the significance of a systemic approach to transformation. Orientation in this journey is vital. Our concept serves as a guiding light, akin to the outer band on a motorway, directing individuals toward their destination.

Ethical considerations are paramount. We must honor individuality and remain receptive to correction. Studies on my work by the Darden Business School introduced the concept of moral imagination—envisaging a better world and fostering conditions for positive innovation. This principle is crucial in both education and business. A few years back, I consciously decided to channel my energies towards those receptive to change rather than expending effort on convincing skeptics. This shift has been more fulfilling and effective, fostering communities that resonate with shared mindsets and values.

Recognizing the importance of focusing on positive energy marked a significant turning point in my journey. Everyone is entitled to their path, but embracing a different understanding is vital. The essence lies in building a community and network capable of making a substantial impact. A pertinent question arises: How do we navigate the learning landscape in an era where knowledge rapidly becomes obsolete due to technological advancements? Continual learning is crucial to stay abreast of changes, presenting opportunities and risks, particularly in maintaining quality amidst swift adaptations.

This book aims to impart knowledge and insights, helping others to make informed decisions in their personal and professional lives. A circular economy is predicated on preserving resources for future generations. Cradle to Cradle® stands at the forefront of this ideology, incorporating design principles that ensure products can be disassembled and reused while maintaining their performance, aesthetics, and material integrity.

Sustainability and circularity fall short in education—they don't ensure resource conservation or non-toxicity. Cradle to Cradle® fills this gap, offering guidance and certification for assurance. In today's interconnected world, collaboration is vital. Open source is a well-known concept, yet many businesses and industries cling to traditional models.

The certification approach bridges this gap, facilitating dissemination without full disclosure. I believe that, over time, business transparency will rise to meet consumer expectations. We're moving towards a future where crucial information is more accessible, aiding in quicker and more efficient product transformation. This evolution is fundamental to our discussions of chemistry, materials, product development, and education.

Stakeholders play a pivotal role in achieving successful outcomes. Defining interfaces and promoting a holistic approach are essential. European policies, for instance, have set boundaries for businesses and industries to prevent resource loss.

Such an interconnected approach is indispensable for genuine change. Cradle to Cradle® is built on three principles. First, waste equals food, advocating for safe products for biological or technical cycles, with the garbage bin radically designed out. Secondly, maximize solar income, aim for 100 percent renewable energy usage, and avoid fossil fuels. Thirdly, fostering diversity in resource use and human coexistence.

These principles are the bedrock of our implementation methodology. Our current focus is on implementation. While these principles have existed for thirty years, understanding and acceptance have been relatively recent. The Reference Model Cradle to Cradle™ has been instrumental in assisting companies to decide whether to innovate radically for the future. Our objective is to provide the necessary information to facilitate these decisions.

Cradle to Cradle® vs. LCA (Life Cycle Assessment)

	Cradle to Cradle ®	LCA – Life Cycle Assessment
What is it?	Innovation framework, Business concept (Circular Economy), but maintaining the quality of raw materials	Method for measuring the environmental impact of products over the entire lifecycle
Philosophy	It is possible to design products with a positive impact on people, the environment and economic profit (the three P's)	„All" products pollute, they all require extraction of raw materials and there is always some form of waste left over
Approach	Eco-effectiveness: developing a product with positive qualities. The process is part of the ultimate goal.	Eco-efficiency: doing more with less. Improving the ratio between economic value and environmental impact. The aim is to measure the result, not the process
Design support	Use the 3 guiding principles to establish a clear direction: waste = food, renewable energy, Respect diversity	Use hot spots to set priorities for improvements.
Environmental impact	Maximization of positive effects on people, their environment and the future availability of high quality raw materials.	LCA is used as a measuring instrument in eco-design, whereby hot spots (life cycle elements with the biggest – negative – environmental impact) are identified so that designers can set priorities for improvements.
Ecological footprint	Develop a positive beneficial footprint.	Measure the footprint and let designers decide how to handle with it.

© epeaswitzerland gmbh

Figure 3: Cradle to Cradle® compared to LCA.

While Cradle to Cradle® represents the circular and LCA the linear mindsets.

Cradle to Cradle® Projects

Reference Model Cradle to Cradle™

1. Defining purpose of **the product**
2. Determining the **metabolism: biological or technical**
3. Definition **closing the loop scenarios**
4. Definition **areas of innovation (chances/risks)**
5. Development of **product criterias and product purposes**
6. Setting priorities of **the criterias**
7. ABC-X categorisation of **the ingredients**
8. Developement of **the positive list**
9. Phase out plan X **(red) substances**
10. Implementation **product design**
11. Implementation **processes production and supply chain**
12. Strategy implementation of **closing the loop scenario**
13. Development **marketing statement** (certification yes/no)
14. Influences **consumer behaviours**
15. Financial **investments**
16. **Influences** consumer behaviours
17. **Marketing** focus
18. After sales services **after the product launch**

Source: epeaswitzerland gmbh 2011

© epeaswitzerland gmbh

Figure 4: Reference Model Cradle to Cradle™. Providing successful guidance and orientation for material health, circularity, radical innovation and business model, is like taking project members on a golf course.

Reference Model Cradle to Cradle™
Supply Chain Domino Knowledge Transformation™

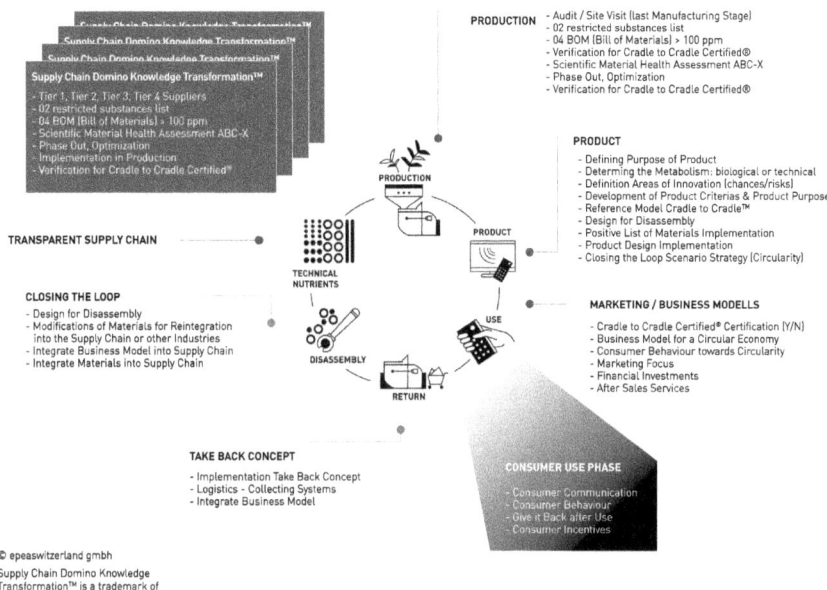

Figure 5: Supply Chain Domino Knowledge Transformation™ applied in the Reference Model Cradle to Cradle™.

The holistic approach illustrates how to implement circular economy and Cradle to Cradle® along the product life cycle.

CHAPTER 6
Practicing Design Innovations

Cooking Book
Cradle to Cradle® Design Innovations

From Rebel to Radical Innovator

Cooking Book Cradle to Cradle® Design Innovations

EDITOR
Albin Kälin, CEO, epeaswitzerland gmbh.

"I'm actually left-handed. When I got to school, I was forced to write on the right. Back then I asked why, but I didn't get an answer. This experience has taught me not to accept anything whose sense I do not understand. I want to shape my life so that it makes sense for me and for the community. In doing it becomes visual how I take responsibility. I work professionally to enable innovations to develop new concepts of production that no longer produce waste, but become 'nutrients' for other products. By showing new ways to the old system, I find my track of interface, where I am neither saying yes nor no and I am confident to bare my responsibility."

Coaching from Linear to Circular Thinking

We have been educated and trained to think linear. All our systems are based on linear design. Just everything. Now we have to learn to think in cycles. This is not an easy task, similar as for a lefthanded person to switch to the righthand. Let's start the process, once it makes "click" you can apply it anywhere. And "click" is not making a cycle from the linear line, because the dustbin still exists, it just has to be designed out, with no compromise. This means all products need a redesign "Rethinking the way we make things".

Picture © Circular Flanders

Picture © Circular Flanders, adopted epeaswitzerland gmbh

Practicing Design Innovations

Cooking Book
Cradle to Cradle® Design Innovations: Become a Product Designer yourself

Picture: 22nd April 2016 Sarah Peter Vogt – FHV – Nacht der Forschung – A-Dornbirn

© epeaswitzerland gmbh

From Rebel to Radical Innovator

Cooking Book
Cradle to Cradle Design® Workshop

1. CHOOSE PRODUCT

Your space to write it down

2. SCENARIO (choose only one)

☐ Biological Cycle

☐ Technical Cycle

3. DEFINITION MATERIALS + CHEMICALS

Your space to write it down

BIOLOGICAL NUTRIENTS	TECHNICAL NUTRIENTS
Materials	
Wood - Packaging - Decor	Metal - Steel, Aluminium, variations
Paper - Cardboard- Packaging	Glass
Filter, Fibers – Technical Textiles, Non Wovens, Wires (monofilament)	Pigments, Chemicals – (within the technical cycle) Masterbatches, Silikones, Flameretardane, Coolant
Plastic – biolodegradable	
Plastics (PLA, Ecoflex, EcovioBASF, Master-Bi Novamont, CA (Celluloseester), PLA (Polylacticacid, Polymilchsäure), PHA (Polyhydroxyalkanoate), PCL (Polycaprolactone), Starkderivates and special Copolyester like PBAT (Polybutylenadipat-terephthalat) and PBS (Polybutylensuccinat)	Plastics - Thermoplast, Duroplast, Elastomere, ThermoplastiElastomeres.
Pigments, - Masterbatches Process Chemicals, Silicones, Flameretardants, Inks, oils, adhesives, Coolant	
Components	
Adhesives	Electronics lead etc.
Plastics	
Cleaning Detergents	Wires – PET, Copper
	Batteries, Circuit Boards, LCD

4. LAYOUT OF THE CYCLES

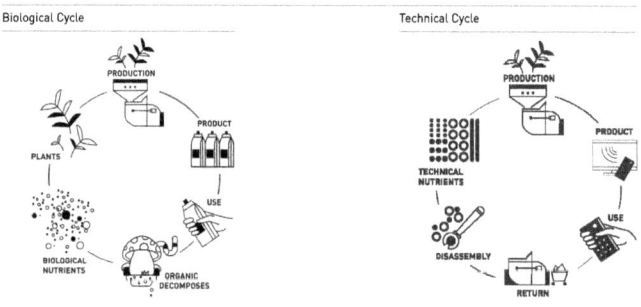

Biological Cycle

Technical Cycle

5. INFLUENCES ON CONSUMER BEHAVIOUR

Your space to write it down

© epeaswitzerland gmbh 2016

Practicing Design Innovations

Cooking Book
Cradle to Cradle Design® Workshop

---

1. CHOOSE PRODUCT

Please select your own product idea based on what you want. e.g. skateboard, toothbrush, milk packaging or any other.

Your space to write it down

Note

Your choice to select a more or less complex product, but try to keep it straight, not to ambitious. The product should not be from your professional experience, it is easier to think out of the box getting into a new field.

© epeaswitzerland gmbh

From Rebel to Radical Innovator

Cooking Book
Cradle to Cradle Design® Workshop

2. SCENARIO

☐ Biological Cycle

☐ Technical Cycle

Your space to write it down

Notes

Important! Investigate both cycle concepts, think them through the whole loop. Are the products safe and cyclable and can you maintain the quality.

Biological means, the product gets back to the earth and becomes nutrients for other organism again, but how to compost? Home or industrial? What happens when toxic ingredients are in the product?

Technical means, the product can be technically cycled in a system and remain the quality properties, being nontoxic. Is the technology available?

Do not mix the two cycles unless the design allows no loss of materials. e.g. textiles, cotton/polyester fibers are a no go, no technology available today to keep both materials in the loop. Many teams are working on this, but no real economical solution is available

© epeaswitzerland gmbh

Practicing Design Innovations

Cooking Book
Cradle to Cradle Design® Workshop

3. DEFINITION MATERIALS + CHEMICALS

	BIOLOGICAL NUTRIENTS	TECHNICAL NUTRIENTS	
	Materials		
☐	Wood – Packaging, Decor	Metal – Steel, Aluminium, variations	☐
☐	Paper, Cardboard – Packaging	Glas	☐
☐	Filters, Fibers – Technical Textiles, Non Wovens, Wires (Monofilament)	Pigments, Chemicals – (within the technical cycle) Masterbatches, Silikones, Flameretardane, cooling agent	☐
☐	Plastic – biolodegradable Plastics (PLA, Ecoflex, EcovioBASF, Master-Bi Novamont, CA (Celluloseesterl, PLA (Polylacticacid), PHA (Polyhydroxyalkanoate), PCL (Polycaprolactone), Starkderivates and special Copolyester like PBAT (Polybutylenadipat-terephthalat) and PBS (Polybutylensuccinat)	Plastics – Thermoplast, Duroplast, Elastomere, ThermoplastiElastomeres	☐
☐	Pigments, Chemicals – Printing inks – Masterbatches Process Chemicals, Silicones, Flameretardants, inks, oils, adhesives, cooling agent		
	Components		
☐	Adhesives	Electronics – scare elements, silicones,	☐
	Plastics, Paper, Labels	lead, etc.	
☐	Cleaning Detergents	Wires – PET, Copper	☐
☐		Batteries, Circuit Boards, LCD	☐

Notes

Please choose the materials, ingredients you need for your product.

Dark-grey *materials are being used for a biological cycle*
Light-grey *materials are being used for a technical cycle*

Reminder: do not forget chemicals, dyes, printing inks, additives, catalyst, masterbatch, galvanic, color coating, lacquer, adhesives, paper coating and others.

© epeaswitzerland gmbh

Cooking Book
Cradle to Cradle Design® Workshop

---

4. LAYOUT OF THE CYCLES

Biological Cycle

Your space to write it down

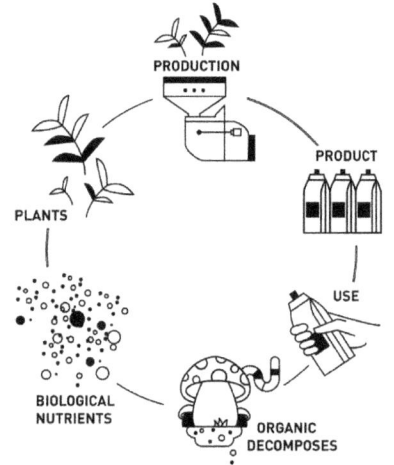

Technical Cycle

Your space to write it down

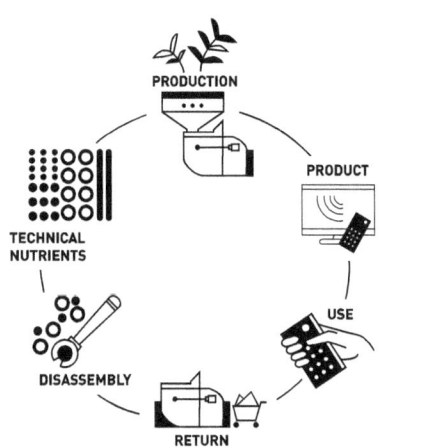

Notes

Draw the cycle concept how to close the loop, from raw material extraction, production, use, return, cyclability and regenerate resources for their next life.

© epeaswitzerland gmbh

Practicing Design Innovations

Cooking Book
Cradle to Cradle Design® Workshop

5. INFLUENCES ON CONSUMER BEHAVIOUR

Notes

To make your concept work, what needs to change for the consumer to close the loop.
An interesting observation with the children taking this exercise. They do not understand the question, when you explain, they react spontaneous. "Just give it back".

© epeaswitzerland gmbh

Cooking Book
Cradle to Cradle Design® Workshop: **REFERENCE MODEL CRADLE TO CRADLE™**

FOR PROFESSIONALS IN BUSINESS

- Defining purpose of **the product**
- Determining the **metabolism: biological or technical**
- Definition **closing the loop scenarios**
- Definition **areas of innovation (chances/risks)**
- Development of **product criterias and product purposes**
- Setting priorities of **the criterias**
- ABC-X categorisation of **the ingredients**
- Developement of **the positive list**
- Phase out plan X **(red) substances**
- Implementation **product design**
- Implementation **processes production and supply chain**
- Strategy implementation of **closing the loop scenario**
- Development **marketing statement** (certification yes/no)
- Influences **consumer behaviours**
- Financial **investments**
- **Influences** business models
- **Marketing** focus
- After sales services **after the product launch**

Source: epeaswitzerland gmbh 2011

Notes

For Professionals in Business this Reference Model is applied and has proven to be successful. It enables a discussion on eyelevel among the participants, decisions can be made on common interest, respecting the values of the individual.

© epeaswitzerland gmbh

CRADLE TO CRADLE® DESIGN INNOVATION
Insights to develop, use, apply healthy materials which are safe and circular for the textile and fashion industry

The textile and the fashion industry are the second most polluting industry, next to the oil industry. Approximately 60% of all the textile production is based on a mixture of fibers (cotton/polyester). In a circular design model, a "no go", as biological and technical cycle materials are mixed. Many teams are working globally on finding a solution for recycling, but can this be economically feasible? Will it pay off? If not, it will not be successful. Are there other design solutions that should be approached? Yes, there are, but this causes a different way of thinking out of the box and creating innovation, for example biodegradable co-polyester.

But toxics are used widely in the textile industry. A few examples illustrating the emergency for action.

Cancerogenic Antimony used as Catalyst in Polyester fibers and PET Packaging

Toxic chemicals are being used in polyester or PET packaging materials. A heavy metal called Antimony, is used as catalyst to produce polyester or PET. Antimony is toxic, cancerogenic and it migrates. What does that mean? There are available alternatives on the market, but industry is refusing to use them. Why are they doing this? Why do they not switch to healthy materials? Industry bear no risk, as they are in legal compliance. Industry has to declare all toxic materials above 1000 ppm (parts per million) 0.1%. The amount of catalyst is app. 200-300 ppm (0.02-0.03%). And for the migration, 40 mg/lt are legally accepted, effectively 6-10 mg/lt are migrating. What does that mean to society after a widely use in 30 – 40 years? And just using recycled PET (rPET) is no solution either, because it contains Antimony.

Membranes made from Polytetrafluoroethylene (PTFE) or called Teflon (Du Pont)

"No go Chemicals" Perfluorocarbons break down within the body and in the environment to PFOA, they are the most persistent synthetic chemicals known to man. Once they are in the body, it takes decades to get them out – assuming you are exposed to no more. Alarmed by the findings from toxicity studies, the EPA announced on December 30, 2009, that PFC's (long-chain perfluorinated chemicals would be on a "chemicals of concern" list and action plans could prompt restrictions on PFC's and the other three chemicals on the list. Polybrominated diphenyl ethers (PBDEs), phthalates and short-chain chlorinated paraffins (SCCPs) Three of these four chemicals are used in textile processing.)

PVC (Polyvinylchloride)

"No go Chemicals" its production, use, and disposal results in the release of toxic, chlorine-based chemicals. These toxins are building up in the water, air and food chain. The result: severe health problems, including cancer, immune system damage, and hormone disruption.

Need for Positive list of Chemicals and Dyes

In dyes and textile chemicals, the presence of sulphur, naphthol, vat dyes, nitrates, acetic acid, soaps, enzymes chromium compounds and heavy metals like copper, arsenic, lead, cadmium, mercury, nickel, and cobalt and certain auxiliary chemicals all collectively make the textile effluent highly toxic.

Compostable, but Is it environmentally safe?
Many compostable products are sold in the market today. How can we be assured that they are none toxic? Using printing inks, polyester buttons, polyester sewing threads and others.

The solution:

Making public available positive lists of materials, chemicals, dyes and trims for the textile and fashion industry according the Cradle to Cradle Certified™ certification product standard. But how to make it accessible for daily use: www.c2ccertified.org have materials, dyes and chemicals in the material health certificate section. Or we can find a brand who will make it online available for the public.

NOTES

Figure 6: A book in a book. Become a designer yourself.

CHAPTER 7

Transforming Design with Cradle to Cradle® Principles

Today's design approaches and processes still tend to follow a philosophy inculcated through decades of modernism and postmodernism. "Design" is still defined by commercial success and efficiency throughout the design stage, particularly in production. Individualism and branding are key criteria. Social design, championed in the 1960s by Victor Papanek and Lucius Burckhardt, enlarged the scope of design with inclusion, social values, and environmental criteria. However, this approach failed to have a widespread influence. In the 1980s, few individuals and NGOs tried to include environmental issues more prominently in the mainstream of design-practice, with some more success during the 1990s, possibly as a variant of the postmodern movement. Social design was revisited only a few years back and has since received fresh attention through numerous expositions. Nonetheless, much of the teaching and practice of product and industrial design today still fall short of a comprehensive definition. The radical shift championed by activists like myself advocate for a design paradigm expanded with all-inclusive approaches—not incrementalism by traditional business and engineering practice, not simply by harming our biosphere less, but in anticipation of a world where our grandchildren can grow up in a healthy environment.

Consider this: products, objects, and/or services are manifestations of a transformative process, involving inputs and outputs of matter and energy. Matter itself is made of material, thus chemical substances and stocked energy. The design of products and/or services embodies the ingenious process to formulate specific relationships between chemical substances and materials: the transformation process, the usefulness and applicability of the outcoming result through specific functions. By standards of modernism, the inherent value of a product and/or service lies in the materials, chemical substances, and the energy input and output over the lifespan of a product and/or service. Likewise, modernism has taught us that the perceived value of a product and/or service depends on the utility as defined by the consumer, user, and owner. Modernism has taught us to think and act linearly within units of times: take, make, waste.

In contrast, Cradle to Cradle® (C2C®) represents a transformative circular concept, aiming to create positive-impact systems and products. It pioneers the integration of sustainability and regeneration into the creative processes of design and economics, marked by the adoption of C2C® principles, which are systematically applied for all product systems across the entire supply chains and all industries, including the textile industry and beyond.

Circular thinking redefines design as a conscientious act of responsibility, based on Nature and counter to reductionism practice. Effectively, uncompromising design begins with the formulation of processes, products and/or services based on "assessed safe chemicals" to yield, not only the targeted utilitarian functionality for a singular use phase, but multiple and consecutive industrial reintegration of material and chemical substances. If this principle of maintaining only pure and safe-assessed chemicals in material streams is violated, then the "circular economy" becomes banalized, downcycling and waste continues, and linear thinking prevails. Circularity and material health by Cradle to Cradle® is uncompromising.

Redefining design, manufacturing, and consumption is more than just following C2C®'s core principles. As my work demonstrates, it is a model for harmonizing design with Nature's cycles, offering inspiration and guidance for future designers, manufacturers, and policymakers dedicated to positive global transformation.

The Birth of Cradle to Cradle® in Design

The Cradle-to-Cradle® design concept stands out in the linear world of "take-make-waste" industrial processes. C2C® rethinks product design and usage by imbedding production into natural cycles. My adoption of C2C® in the textile industry represented a crucial shift towards sustainable design, blending innovation with environmental responsibility.

Applying C2C® Principles

C2C® embodies five core principles: (1) Material Health, (2) Product Circularity, (3) Clean Air and Climate Protection, (4) Water and Soil Stewardship, and (5) Social Fairness. I have seen these principles come to life in various industries, setting new sustainability standards.

Broader Impact and Future Prospects

C2C®'s influence extends beyond environmental benefits, influencing economic models, consumer behavior, and policies. We assess the broader implications of my work and the C2C® movement, considering future innovations and technology's role in advancing sustainable design and manufacturing.

We conclude by reaffirming the transformative power of Cradle to Cradle® design as both a philosophy and a practical tool for achieving fully integrated sustainability-design. My legacy exemplifies the impact of visionary action in guiding design evolution toward a regenerative future.

CHAPTER 8

Transforming Science

Material Health ABC-X Methodology

In sustainable innovation, transforming science is as much about pioneering new pathways as it is about revisiting the fundamentals with a fresh perspective. Science is also based on linear thinking and applies a Cradle to Grave approach, where waste is, to a large extent, discarded. To transform science is an adventure with many hurdles to overcome.

The key element is to include worst-case scenarios of exposure to the environment. Another issue is how much data to include. Science and legislation are based on the disclosure of toxic ingredients: for example, more than 1000 ppm (parts per million) (0.1%) are declared in the Safety Data Sheets (SDS) in products or materials. The Cradle to Cradle® ABC-X Methodology is focusing on full disclosure: less than 100 ppm (< 0.01%). The Material Health ABC-X Methodology of a classification tool stands at the forefront of this transformative approach, offering a blueprint for evaluating and classifying materials for their utility and their impact on health and the environment. But there is always a "but." How can this scientific concept find global acceptance when greeted with widespread skepticism? Through the implementation of the Cradle to Cradle Certified® standard, this could be accomplished.

ABC-X Categorisation

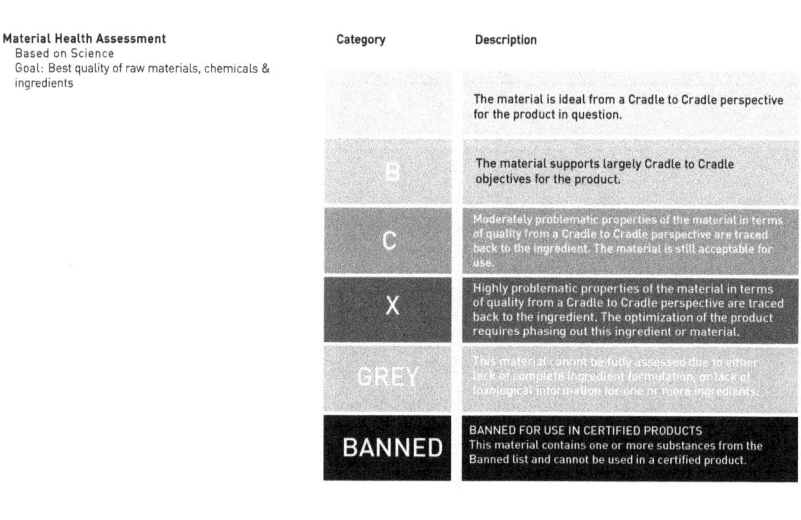

Figure 7: Material Health with ABC-X Categorization.

This assessment methodology brings about sustainable innovation and transforms science.

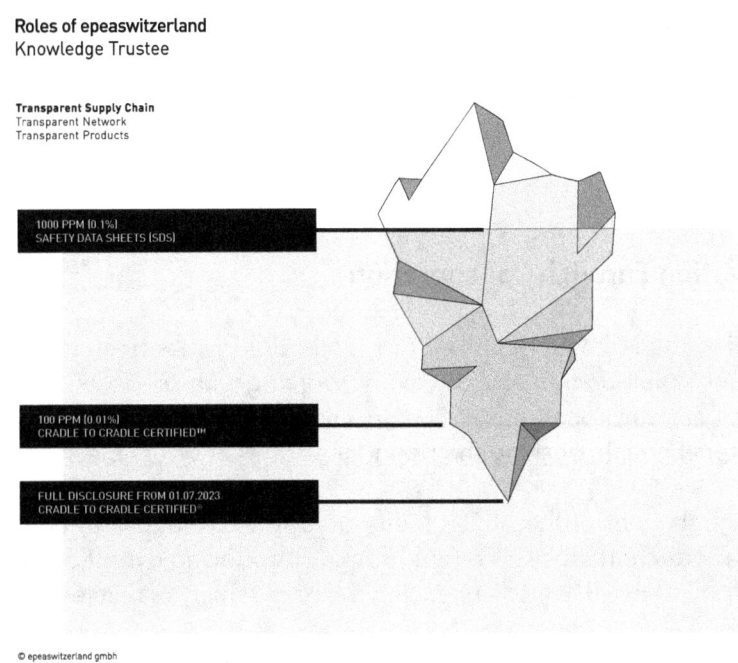

Figure 8: Full transparency and full confidentiality.

Achieving full transparency going beyond common limits while gathering and protecting secret industry information is the role of Knowledge Trustee.

A Personal Journey Towards Understanding

My journey toward understanding and applying the Material Health ABC-X methodology began in the early stages of my career amidst the lush landscapes of Switzerland, where the interplay of nature and industry provided a constant reminder of our responsibility towards the environment. I quickly realized that true innovation in the textile industry wasn't just about creating fabrics that looked good and performed well—it was about ensuring these materials safeguarded our health and the planet

The ABC-Xs of Material Health

The ABC-X Methodology isn't just a classification system—it's a philosophy that challenges us to think deeply about the materials we use daily.

Innovation Through Collaboration

Applying the Material Health ABC-X methodology has been a journey involving collaboration and discovery. Working with scientists, industry experts, and environmentalists, we've navigated the complex landscape of material health, pushing the boundaries of what's possible.

Through the lens of the ABC-X Methodology, we've transformed traditional production processes—for example, introducing dyes that reduce water pollution while also enhancing the recyclability of textiles.

We've collaborated with partners to develop new composite materials that marry the strength and durability required for high-performance usage with the safety and sustainability of materials that meet the rigorous ratings of ABC-X.

Our journey has also taken us into the realm of product innovation, where we've reimagined product lifecycles, creating circular systems that restore rather than deplete.

Looking Ahead: The Future of Material Science

With each material we assess, classify, and innovate, we're not just transforming products—we're transforming our relationship with the planet.

The science of materials is evolving, driven by a collective determination to do better and be better. The Material Health ABC-X Methodology demonstrates the power of science wielded with conscience and courage.

In the fabric of our shared future, let sustainability be the thread that binds us, innovation the pattern that guides us, and the health of our planet the canvas upon which we create.

Embracing Challenges: The Journey to Green Chemistry

Pursuing green chemistry within the textile industry has been a professional and personal crusade for me. The Material Health ABC-X Methodology offered a framework and a lens through which to view the entire lifecycle of products—from raw material extraction to end-of-life recycling or decomposition. This method has underscored the importance of selecting materials that benefit our immediate needs while preserving long-term ecological health.

My awakening to the critical need for sustainable materials came during a visit to an Asian manufacturing plant. There, amidst the buzz of machinery and the relentless pace of production, I witnessed firsthand the consequences of neglecting material health: polluted waterways, compromised worker safety, and communities living in the shadow of industrial waste. This experience was a stark reminder of the industry's impact and the urgent need for change. It propelled me to advocate for the ABC-X Methodology not just as a guideline, but as a mandate for innovation and responsibility.

Many milestones of innovation and collaboration have marked the journey since then. For instance, transitioning ABC-X positive list materials to safer alternatives required technological innovation and a shift in mindset. It led to the development of a new class of fibers, safe for biological cycles, which could be recycled or safely returned to the earth.

One of our most significant achievements was the introduction of a fabric coating derived from plant-based polymers. This breakthrough, which began as an experiment in a lab, resulted in a water-repellent, durable, and fully biodegradable product—an authentic ABC-X positive list material. This innovation demonstrated the feasibility of sustainable alternatives and set a new standard for the industry.

Adopting the Material Health ABC-X Methodology has sparked a ripple effect across the textile industry and many other industries, encouraging companies to scrutinize their materials and processes through the lens of sustainability. It has fostered partnerships with environmental organizations, research institutions, and competitors, united by a common goal to mitigate the industry's ecological footprint.

As we look to the future, many sustainability challenges remain, but the successes of the past illuminate the path forward. The Material Health ABC-X Methodology is not just a tool for classification but a compass guiding us toward a more sustainable and equitable industry. It embodies the principle that environmental stewardship and economic prosperity can go hand in hand, driving innovation that benefits our generation and generations to come.

CHAPTER 9

Cradle to Cradle Certified® Certification

From Linear Certification Schemes to Circular for all Products of all Industries

In the journey toward sustainability and circularity, the transformation of certification standards is most crucial. The Cradle to Cradle Certified® certification is an alternative and a revolutionary paradigm shift from linear to circular for products across all industries. This shift challenges and redefines traditional certification schemes, embracing a comprehensive life cycle approach that aligns with the principles of the circular economy.

Life Cycle Assessment (LCA), Microfibers, and Bottles in the Sea

The urgency for this transformation is underscored by the growing environmental concerns associated with linear product life cycles. Although valuable, traditional Life Cycle Assessments (LCAs) have often focused on minimizing negative impacts, rather than promoting unconditional positive environmental and health outcomes. The issue of microfibers in our oceans and the ubiquitous presence of plastic bottles are emblematic of a broader systemic failure. These problems are not merely about pollution—they reflect an inherently unsustainable design and production philosophy. The linear "take-make-dispose" model has led to a crisis of accumulation, where the durability of materials like plastics becomes a liability rather than an asset.

The Cradle to Cradle Certified® approach offers a radical departure from this model. By reimagining the entire lifecycle of products, it aims not just to mitigate harm but to foster a regenerative system where materials that are perpetually cycled remain in good quality and are non-toxic.

Label Comparison

Due Diligence Supply Chain & Due Diligence Risks Assessment in Supply Chain

© epeaswitzerland gmbh

- ● Meets Criteria
- ○ Complies partially the criteria
- X Does not meet criteria

	CRITERIA	Material Health	Product Circularity	Clean Air & Climate Protection	Water & Soil Stewardship	Social Fairness	Indoor Airquality	Industry Specific Certification
	CERTIFICATIONS							
Products	Cradle to Cradle Certified®	●	●	●	●	●	●	●
	EU-ECO Label	O	X	O	O	X	X	X
	Blauer Engel	O	X	O	X	X	X	X
	PEF / LCA (Product Environmental Footprint)	O	X	O	O	X	X	O
Textiles	California Proposition 65	O	X	X	X	X	X	X
	Bluesign	O	X	X	X	X	X	X
	Step Made in Green	O	X	X	X	●	X	X
	Oeko Tex	O	X	X	X	X	X	X
Organic	Bio EU-Öko	O	X	X	●	X	X	X
	FSC	O	X	X	●	X	X	X
	GOTS	O	X	X	●	X	X	X
	IVN	O	X	X	●	X	X	X
	UTZ	O	X	X	●	X	X	X
	Rainforest Alliance	O	X	X	●	X	X	X
	RED cert	O	X	X	●	X	X	X
Social	SeedGuard	O	X	X	●	X	X	X
	SA8000	X	X	X	X	●	X	X
	Fairtrade	X	X	X	X	●	X	X
	Grüner Knopf	X	X	X	X	●	X	X
ISO Management Systems	RBA Verhaltens-kodex	X	X	X	X	●	X	X
	ISO 14001	X	X	●	X	X	X	X
	ISO 26000	X	X	X	X	●	X	X
Building	ISO 45001	X	X	X	X	●	X	X
	SCC/SCP	X	X	X	X	●	X	X
	DGNB	X	X	O	O	X	O	X
	LEED	O	X	O	O	X	O	X
	BREEAM	O	X	O	O	X	O	X

	CRITERIA	Human Rights	Environmental Compliance
	DUE DILIGENCE SUPPLY CHAIN		
Products	Cradle to Cradle Certified®	●	●
	EU-ECO Label	X	X
	Blauer Engel	X	X
	PEF / LCA (Product Environmental Footprint)	X	X
Textiles	California Proposition 65	X	X
	Bluesign	X	X
	Step Made in Green	●	●
	Oeko Tex	X	X
Organic	Bio EU-Öko	O	●
	FSC	●	●
	GOTS	●	●
	IVN	O	●
	UTZ	O	●
	Rainforest Alliance	O	●
	RED cert	O	●
Social	SeedGuard	O	●
	SA8000	●	X
	Fairtrade	●	X
	Grüner Knopf	●	X
ISO Management Systems	RBA Verhaltens-kodex	●	X
	ISO 14001	X	●
	ISO 26000	O	X
Building	ISO 45001	O	X
	SCC/SCP	O	X
	DGNB	X	X
	LEED	X	X
	BREEAM	X	X

CRITERIA	Renewable Resources	Petro Chemical Resources	Chemicals	Pigments	Polymers	MATERIALS	Wood	Metals	Plastics renewable resources	Plastics petro chemicals	Fibers renewable resources	Fibers petro chemicals	Rubber
DUE DILIGENCE RISK ASSESSMENT IN THE SUPPLY CHAIN													
Water Quantity	●	●	●	●	●		●	●	●	●	●	●	O
Water Quality	●	●	●	●	●		●	●	●	●	●	●	O
Soil Quantity	●	●	O	O	O		●	●	●	●	●	●	O
Soil Quality	●	●	O	O	O		●	●	●	●	●	●	O
Sustainable Agriculture / Forestry	X	X	X	X	X		X	X	X	X	X	X	X
Plantages Agriculture / Forestry	●	X	O	O	O		●	X	●	X	●	X	●
Foodstock	●	X	O	O	O		●	X	●	X	●	X	●
Extraction / Mining	X	●	O	O	O		X	●	X	●	X	●	X
Restricted Substances Chemicals Risk	●	●	O	O	O		●	●	●	●	●	●	●
Toxic Process Chemicals Risks	●	●	O	O	O		●	●	●	●	●	●	●
Toxic Product Chemicals Risks	●	O	O	O	O		●	O	●	O	●	O	●
Impurities Risks in chemical production	X	O	O	O	O		X	O	X	O	X	O	X
Contamination Risks Processes	●	X	O	O	O		●	X	●	X	●	X	●
Contamination Risks Products	●	X	O	O	O		●	X	●	X	●	X	●

Figure 9: Label comparison.

The label jungle is confusing and a challenge for consumers. All labels are based linear, with the only exception of Cradle to Cradle Certified® designed for circularity.

Founding an Independent, Not-for-Profit, Global Certification Product Standard for a Safe and Circular Future

Establishing an independent, not-for-profit global certification product standard marked a significant milestone on this journey. The Cradle to Cradle® Products Innovation Institute, steward of the Cradle to Cradle Certified® certification, embodies the commitment to a safe and circular future. The Institute was the outcome of a collaborative effort by visionaries, industry leaders, and environmental advocates who recognized the need for a certification that transcends traditional boundaries and sectors.

As previously noted, the Cradle to Cradle Certified® certification evaluates products across five critical performance categories:

1. Material Health: ensuring materials are safe for humans and the environment.
2. Product Circularity: enabling a circular economy through regenerative products and process design.
3. Clean Air & Climate Protection: protecting clean air, promoting renewable energy, and reducing harmful emissions.
4. Water & Soil Stewardship: safeguarding clean water and healthy soils.
5. Social Fairness: respecting human rights and contributing to a fair and equitable society.[11]

One of the significant challenges is the need for transparency and collaboration across the supply chain. Achieving Cradle to Cradle Certified® certification requires detailed knowledge of the materials and chemicals used in products, necessitating closer relationships between

---

[11] "Media Kit", Cradle to Cradle Products Innovation Institute, Accessed June 4, 2024, https://api.c2ccertified.org/assets/c2cpii-media-kit-april-2023.pdf

manufacturers, suppliers, and assessors. Moreover, the shift towards clean air and climate protection poses logistical and technological hurdles, particularly for industries reliant on high-energy processes.

C2C Certified Material Health Certificate™
Cradle to Cradle Certified® Certification

Cradle to Cradle Certified® is a registered trademark of C2CPII
www.c2ccertified.org

Figure 10: Cradle to Cradle Certified® and C2C Certified Material Health Certificate™ certification scheme. This comprehensive approach sets a new standard for product certification and drives innovation and improvement in product design manufacturing, consumption, and reintegration into the supply chain. By integrating these principles and implementing the Cradle to Cradle Certified® Certification across industries, businesses can reduce their environmental footprint liabilities risk while unlocking new opportunities for growth and competitiveness in a circular economy. Companies must rethink their product design, supply chain management, manufacturing processes, take back systems, and business models. This transformation demands a shift from viewing sustainability as a compliance issue to seeing it as a core driver of innovation and value creation.

Despite these challenges, the adoption of Cradle to Cradle Certified® principles offers profound benefits. Companies that embrace this approach can enhance their brand reputation, meet the growing consumer demand for sustainable products, and contribute to a healthier biosphere. Moreover, the certification process itself can uncover efficiencies and innovations that lead to cost savings and new market opportunities.

A Global Movement for Change

The Cradle to Cradle Certified® certification is more than a set of criteria for product evaluation. It is a movement toward a new industrial revolution. This revolution is characterized by an economy where growth is decoupled from environmental degradation, where products are designed with their next use in mind, and where all materials are viewed as resources for future generations.

The journey towards a safe and circular future is complex and requires the collective effort of all stakeholders. Governments, businesses, and consumers must work together to create the conditions for success. This includes developing supportive policies, investing in research and development, and fostering an innovation and regenerative culture.

As we stand at the crossroads of history, the choice is clear. We can continue toward linear consumption, with its inevitable consequences of pollution, waste, and environmental degradation. Or we can choose a different route that leads to a world where materials flow in closed loops, energy comes from renewable sources, and products are designed to enhance the health of people and the planet. The Cradle to Cradle Certified® certification offers a blueprint for transforming certification and, with it, the fabric of our economy.

Innovations in Material Health and Reutilization

A cornerstone of the Cradle to Cradle Certified® certification is its emphasis on material health, ensuring that products are composed of materials safe for human health and the biosphere. This focus has spurred innovation in material science, leading to the development of new non-toxic materials and the reevaluation of traditional ones. For instance, companies are now producing biodegradable plastics and non-toxic dyes, transforming industries from textiles, plastics to packaging and other industry segments.[12]

Similarly, the principle of product circularity has led to innovative business models that prioritize durability, repairability, and recyclability. Companies are exploring leasing models for everything from clothing to electronics, ensuring that products are returned, disassembled, and their materials reused or recycled in defined streams. This reduces waste, strengthens customer relationships, and opens new revenue streams. The goal, however, is to eliminate the concept of waste.

Overcoming Barriers to Circular Transformation

One significant barrier is the current recycling and waste management infrastructure, which is often ill-equipped to handle and sorting the diversity of materials and products whether designed only for downcycling or real circularity according C2C®. There's a pressing need for investment in new technologies and systems that can efficiently sort, process, and repurpose materials at the end of their use phase.

Moreover, the transition to renewable energy sources and the management of carbon emissions in the manufacturing process require systemic changes within industries and economies. This includes the adoption of green energy technologies, the development of carbon capture and storage solutions, and the redesign of supply chains to minimize carbon footprints.

The Role of Policy and Regulation

Government policy and regulation play a crucial role in facilitating the transition to a circular economy. Incentives for sustainable design, stricter regulations on waste and pollution, and support for green technologies

---

[12] "Cradle to Cradle Certified®", C2C Platform, Accessed June 4, 2024, https://www.c2cplatform.eu/c2c-certified/

can accelerate the adoption of Cradle to Cradle® principles. The EU's Circular Economy Action Plan and the new European Green Deal are examples of policy frameworks that aim to create a more sustainable and resilient economy by promoting circular practices and reducing carbon, ecological, and chemical footprints.

A Vision for a Circular World

The transformation of certification for safe and circular products, as epitomized by the Cradle to Cradle Certified® certification, represents a beacon of hope in our quest for sustainability. It offers a tangible path forward for companies seeking to align their operations with the principles of the circular economy. More importantly, it signifies a shift in our collective mindset—from viewing the environment as a resource to be exploited to seeing it as life supporting system to be cherished, regenerated, and for enabling resilience.

CHAPTER 10

Transforming Products – Safe and Circular Products - Case Studies

Case Studies: Textiles, Home Textiles, Workwear, Fashion, Lifestyle Fashion, Footwear

Textiles

Home Textiles

Essentials: A Patented Function for a Climate Control Seating

Patented in 1987 in the US, Europe, and Switzerland, Climatex® fabrics offered a patented function, "Climate Control Seating," based on the properties of the raw materials (polyester, wool, and ramie). Just the inherited properties of the raw material, without adding any chemicals in finishing, enabled an unlimited function over the entire product lifetime. Transpiration from sitting for many hours on seating can cause an uncomfortable and unpleasant wet sensation. A well-known chair manufacturer at the time offered seating with mechanical ventilators to absorb the humidity. With Climatex®, the wool absorbs humidity and blended with ramie, a hollow natural bast fiber that was used by the Egyptians to wrap in their mummies, wicks away humidity like in a "tunnel." Polyester yarns did not absorb much humidity but can transport it easily and evaporate in a natural way. Climatex® was positioned to achieve market success as a functionable, comfortable textile perfect for seating. The product got a lot of kudos and was competing successfully in the market, but a couple of years later, major design issues were discovered. These required a radical redesign of materials, chemicals, and colors to achieve material health and product circularity.

Essentials: The Worldwide First Cradle to Cradle® Product

William McDonough visited Rohner Textil AG in the fall of 1992 and shared the vision of Cradle to Cradle®. A redesign of the Climatex® fabrics seemed to be the best option to make the first worldwide Cradle to Cradle® Product. Polyester, a material known for its high abrasion resistance, which is a key property for upholstery textiles, needed to be replaced by a material safe for biological cycles. One option was a wool/ramie blend, which was a risky step. Additionally, the color dyes needed a total reset. Michael Braungart and his team of EPEA Int. Umweltforschung GmbH were hired to introduce the scientific Cradle to Cradle® approach. A unique multidisciplinary team, each sharing the same values that motivated them to do this project, successfully enabled this high-risk transformation and innovation adventure. The result: Climatex® Lifecycle™, the first worldwide Cradle to Cradle® product.

Essentials: The Innovation Flame Retardant and Safe for Biological Cycles

Resting on our laurels was not an option. Out of the blue, flame regulations for contract and public installations were tightened because of the fire in Germany at Düsseldorf's Airport that I mentioned earlier. In 1999, Climatex® LifeguardFR™ innovation was born, made from a blend of wool and an inherited flame-retardant viscose, which needed a "redesign for nature" process to become compliant with the Cradle to Cradle® Design Innovation Protocol. The innovation enabled to widen the market segment to include mobility in public transportation. As a result, large players in the business began to see this small company as a threat to their market share. Rohner Textil AG was sold in 1999 to a competitor that had 65 percent of world market share for seating fabrics in aviation.

Another Swiss textile company acquired the Climatex® patents and technology, leading to the formation of Climatex AG.

Case studies on Rohner Textil AG for business leaders are available at IMD Business School, Switzerland 1999 and Darden School of Business University of Virginia 1997.

Climatex AG: Climatex® is the revolution of the textile sector. The company was formed from Gessner, a traditional Swiss company that had been writing textile history for more than 175 years with the courage to drive forward sustainable textile innovation.

The Climatex® brand stands for climatizing and recyclable textiles. They have impressive characteristics and conserve valuable resources. The individual textile components can be separated and reused—a pioneering achievement in the industry.

The patented Climatex® technology revolutionizes the weaving process. The concept produces recyclable fabrics with outstanding properties. They equalize temperatures, regulate moisture, are non-toxic, and very durable.[13]

Essentials: PET Flame Retardant and Safe for Technical Cycles

For public contract installations of interior home textiles, flame retardant regulations must be met. The environmental attributes of flame-retardant chemicals are extremely challenging in a Cradle to Cradle® scenario. A chemical yarn manufacturer was able to innovate and come up with a solution, but without a workable alternative for an environmentally friendly catalyst, only a bronze level Cradle to Cradle Certified® could be awarded. Just replacing one chemical with another does not always work. Criticism and denial challenged the project team for many years. Nonetheless, trials to replace antimony as a catalyst failed over

---

[13] "Climatex Circular Textile Technologies, Climatex, Accessed June 3, 2024, https://www.climatex.com/en/company/

and over. Then, out of nowhere, the yarn manufacturer suddenly was able to offer a flame-retardant solution without antimony. It was a relief to realize that this innovation was created through tirelessly discussing issues, making them visible, and insisting on solutions. Pfister immediately recognized the potential in the contract market and took action, asking the textile supply chain manufacturer to produce a new range of drapery fabrics for the technical cycle, which led to a gold standard Cradle to Cradle Certified® certification. Launched in early 2020, Pfister is still the only company able to offer this product solution. Why has no copy-and-paste from competitors occurred so far? The antimony-free yarns are sold as ordinary goods to the market, but no other competitors seem interested in adopting the right chemicals and colors to go that extra mile. Consumers are still not sensitized to the dangers of antimony as they should be. So, how do we disseminate this awareness on a global scale?

Beyond Limits

Möbel Pfister AG, the parent company of the Switzerland-based Pfister Vorhang Service AG, is the retail market leader in furniture and home interiors with twenty department stores and twelve hundred employees. Cradle to Cradle® is a success story at Pfister. Since the launch of the world's first curtains safe for biological systems in autumn 2017, the range has grown steadily. In 2019, bed linen and terry towels with a Cradle to Cradle Certified® gold certificate were launched.

Essentials: Turkiye Organic Cotton No GMO, Integrated Supply Chain, Closing the Loop

Cotton fabric has had political, economic, social, and industrial impacts throughout history. Cotton has been spun, woven, and dyed for thousands of years. It was used in ancient India, Egypt, and China. In the mid-19th century, cotton comprised over 50 percent of exports from the United States. During the industrialization of agriculture, the environmental impact of growing cotton grew tremendously, with the crop using 2.5 percent of agricultural land. The water usage to grow cotton increased significantly. Today, 23 percent of global pesticide-production goes to cotton culturing, while 75 percent of today's cotton seeds are genetically modified (GMO). Only Tanzania in Africa and Turkiye prohibit its use.

Less than 2 percent of the global cotton market is offered as organic cotton. Can this ever be perceived as sustainable? And yet consumers are demanding organic cotton, with the consequences that retailers are declaring a large group of fashion products to be organic. Fraudulently certified declarations harm the organic image and skepticism feeds distrust into entire industries in agriculture, textile, and apparel.

How can one get out of this dilemma and create substance, content, and transparency? Stefan Grabher, the owner of Mary Rose GmbH in Austria, is convinced that cotton is inherently a gorgeous, precious material. But how to search and find sources of credible and sound raw materials, to build an entire supply chain in the textile and garment industry that can live up to an elevated level of ethics? This adventurous route finally found its destination in Turkiye. For over two millennia, farmers in Turkiye cultivated cotton in small- to medium-sized fields, hence conserving water usage in the region. No pesticides are used for organic cotton farming. The textile industry and cut-and-sew manufacturing have a long tradition in Turkiye. The entire supply chain for the Mary Rose GmbH products is located in an area of one hundred square kilometers. Their home textiles products, from bed covers, terry towels, draperies, cushions and more, embody both durability and superb aesthetics. The products are gold level Cradle to Cradle Certified®.

Home textiles from Mary Rose in Austria are designed to be sustainable with a transparent supply chain, starting with organic raw materials, fair working conditions, compliance with environmental and social standards along the entire supply chain, up to the pollutant-free biodegradability of products. Sustainable business requires building global networks and implementing an integrated, comprehensive approach.

Work Wear

Essentials: Performance Textiles for Workwear, Industrial Washing

Most woven or knitted workwear textiles are a blend of cotton/polyester (CO/PET) to enable wearability and industrial washability. No matter how many technologies are being developed to solve design problems with a repair approach, as long as the system itself is wrong, it remains a no-go in circularity and Cradle to Cradle®.

One way out is the integration of innovative fibers, vinatur® or naNea. This is the alternative to polyester, enabling the safe for biological cycles Cradle to Cradle® design concept. Together with Cradle to Cradle® chemicals, dyes and trims, and workwear apparel are Cradle to Cradle Certified® at gold or silver level. Workwear products are worn in healthcare, as security or governmental uniforms, and to represent corporate identity with unique visibility. Often, a classical reuse or recycling approach is not desired, due to misuse or fraud, with a biological closing the loop being the preferred option.

Inquiries for the highly competitive business need to be placed within public or private tenders. How should innovations and circularity in a tender be promoted within a linear, conservative market? This is a huge obstacle to overcome. It took years to acquire the first projects, and it remains a huge challenge. Only committed pioneers are willing to go the extra mile.

Service providers for industrial washing solutions for workwear play a crucial role within a circular business model, since they are best placed to separate damaged clothes after washing to reintegrate into a closed loop scenario.

This case study illustrates the challenges for innovations in materials to enter markets, products, business models, take-back systems and reintegration resources back into the supply chain.

Essentials: Performance Textiles for Healthcare, Industrial Washing, Take Back

One of the breakthrough projects for circular workwear for a large healthcare clinic in Germany is still confidential at the time of writing, but hopefully the public will soon hear about this advance. The healthcare market is crucial if a society is to save lives while under severe operational and economic pressure, whether this is due to the challenge of getting people to become doctors and nurses or because of the enormous, contaminated waste streams the system generates from products that are not designed to integrate circularity.

In a pilot project, the innovative circular workwear products were evaluated during wear by clinic employees of the dermatology department in their daytime or night shifts. The employees reported high comfort, and even feelings of well-being. Karl Dieckhoff GmbH & Co KG in Germany acquired the project and will deliver the products in 2024. After that, this market segment will be ready for dissemination.

The company is based in Wuppertal, Germany. Dieckhoff is the market leader in offering textile systems in textiles, apparel, and applications for the healthcare and hotel sector.

Services include design, production, consulting, and logistics. Customers include clinics, nursing homes, textile service providers, industrial washing laundries and the hotel and catering industry. The products are Cradle to Cradle Certified® silver level.

Fashion

Essentials: The First Product in Circular Fashion – Starting with a Simple T-Shirt

For the past few decades, textile or clothing companies in Europe have struggled to stay afloat. Few have succeeded, especially in the realm of T-shirts. Trigema in Germany survived to become an icon in this business.

The company was founded in 1919 and now has 1,200 employees. Trigema W. Grupp KG is Germany's largest manufacturer of sportswear and casual fashion. The company is a vertically integrated textile circular knitting, dyeing, finishing, and apparel manufacturing company supplying top-quality sports and leisurewear, as well as underwear and nightwear for women, men, and children. In 2024, the fourth generation, Bonita and Wolfgang Junior Grupp, took the helm of the company. Wolfgang Grupp senior, the former charismatic owner and CEO from 1969 to 2023, positioned Trigema for state-of-the-art technology, social and economic responsibility, and the preservation of Germany as a production location.[14]

Manufacturing clothes in an environmentally conscious manner has always been Trigema's focus. In cooperation with the environmental institute EPEA Int. Umweltforschung Gmbh based in Hamburg, they worked with Michael Braungart and myself from 2005 to 2006. Trigema developed the first biodegradable T-shirt, according to Cradle to Cradle® standards. The collection has since been expanded and is available online and in Trigema stores across Germany.

This T-shirt was promoted "100 percent made in Germany" and exhibited in the German Pavilion at the Expo in Shanghai, China in 2010. The employees of the pavilion were dressed in Trigema Cradle to Cradle® shirts and, at the end of the Expo, the shirts were buried on the premises in the spirit of the Cradle to Cradle® principles. The biodegradable collection, made of 100 percent organic and GOTS-certified cotton, is sold under the brand name Trigema Change. This Trigema Change collection was presented at the Green-Fashion Week in Berlin in 2012. Apart of the Cradle-to-Cradle® guidelines, Trigema also follows a strategy to expand the use of renewable energy. Since 2008, Trigema uses solar panels on roofs for its own energy usage. Trigema Change products are Cradle to Cradle Certified® at gold level.

---

[14] "Trigema Clothing", Anakin, Accessed June 4, 2024, https://www.anakin.co/project/trigema-clothing

Essentials: Luxury, the Masterpiece: Biological and Technical Cycles at Cradle to Cradle Certified® Gold Level, Closing the Loop

Can luxury and sustainability become a symbiosis or even grow beyond that? The Wolford project received funding from the FFG (Oesterreichische Forschungsfoerderungsgesellschaft GmbH) in Austria to start the project in 2014 with a network of fifteen consortium partners and a supply chain within a radius of thirty square miles. epeaswitzerland was integrated as a third party in the role as knowledge and innovation trustee and accredited assessor for the Cradle to Cradle Certified® certification.

Founded in 1950 in Bregenz, Austria, Wolford is the market leader in high-quality skin wear. The company has produced numerous product innovations, some of which are still cutting-edge. Wolford is represented by 229 mono-brand stores and over 2,500 retail partners worldwide in around forty-five countries.

Listed on the stock exchanges in Vienna, Frankfurt, and New York, Wolford is part of the global luxury fashion Lanvin Group, founded by Wolford's Chinese majority shareholder, Fosun.

Luxury hosiery and stockings are the fanciest, most glamour appealing fashion accessory products one can imagine, requiring state-of-the-art technology, finishing processes and high-tech materials. Is there any chance to make them "green" without greenwashing? Wolford, as a global leading brand, was enthusiastic and willing to tackle the challenge.

Wolford manufactures its products solely in Europe, complying with the strictest ecological and sustainable standards. It creates its designs in Milan, Italy. It produces its creations at its headquarters in Bregenz, Austria near Lake Constance, and in the Slovenian town of Murska Sobota. Both regions have a long history and tradition of textile production.[15]

---

[15] "Wolford Company", Wolford AG, Accessed June 3, 2024, https://company.wolford.com/wolford-company/

This particular project took five years. Not until the very end of the project was Wolford able to give a verdict. Five years is very long time to keep an ambitious team of fifteen network partners motivated, especially since some of them were competitors. epeaswitzerland, in its defined independent role, was able to protect knowledge and intellectual property in this business environment and to manage confidentiality in a professional and respectful attitude, while enabling innovation to succeed.

I'll give one example of the many challenges that popped up during the project. The dyehouse manager, who had worked for four decades at Wolford, commented: "In this project, I had to learn everything from scratch, like when I started as an apprentice at Wolford over forty years ago."

Other dimensions of the project included a consumer survey in the US, France, Germany, and Austria, which involved a total of 100,000 respondents. This was a project managed by the Wolford marketing team, an educational institution in Austria, and epeaswitzerland. Wolford Marketing obtained valuable information about consumer needs on sustainability, transparency, material health, saving resources and required incentives to do something positive in returning used consumer goods to the brand.

Taking back worn-out products and close-the-loop was integrated in the project assignments. For the technical cycle, the chemical recycling concept of Polyamide 6 Econyl from Aquafil was used to close the loop. For the biological cycle, a pilot project was undertaken to prove the economic feasibility of industrial biogas/biomass for textiles, fashion, and plastics that were defined to be safe for biological systems. The economics were proven viable with this innovative approach, but authorities have continued to deny official approval up to 2024, since this is considered waste instead of resources. In November 2023, the EU Commission provided information on the biological closing-the-loop concepts, but more research and innovation are required.

Finalizing the project, Wolford was able to approve and communicate the incredible results and decided to integrate the development of luxury products into the collection.

Wolford is the first company in the textile industry to receive Cradle to Cradle Certified® gold level certification for developing environmentally compatible products in both categories: biological and technical cycles.

Fulfilling the highest environmental standards within its own production operations, Wolford's current strategic focus is implementing a sustainability strategy pertaining to the materials it uses for manufacturing purposes. The goal is for 50 percent of all Wolford products to be recyclable by 2025–i.e., either safe for biological cycles or by technical cycles.

Wolford has also developed a new packaging concept that was presented for the first time in 2020 with the summer fashion collection, using gold level Cradle to Cradle Certified® cardboard instead of the customary plastic poly bags for its tights and stockings.

Essentials: Red Socks are only for Devils, Project in Developing Countries

In 2012, a project funded by the Swiss government sought to implement the Cradle to Cradle® innovation design concept in Indonesia's entire textile and fashion industry. A dozen companies were analyzed to demonstrate the concept's feasibility, resulting in positive findings. Unfortunately, the implementation project was not funded and could not be continued. However, PT. Kahatex, one of the companies studied, decided to continue the process. Known as a world-class, fully integrated vertical textile setup, the company was founded in 1979 as a knitting mill and dye house. Covering a combined area of more than 185 hectares in three production sites with a workforce of more than fifty-five thousand, PT. Kahatex is one of the largest family-run textile enterprises in Southeast Asia and directly develops and produces for renowned global brands.

More than 40 percent of the PT. Kahatex products are directly exported to over eighty countries worldwide.

Following the 2012 study, the company pursued a socks project that achieved Cradle to Cradle Certified® certification at the bronze level.

Unfortunately, certification was discontinued in 2023 because of a lack of interest by the same global brands the company produces for.

Lifestyle Fashion

Essentials: The Dream Project, the First Environmentally Friendly, Circular Concept for Lifestyle Fashion

The apparel company Napapijri, part of the VF Corporation, communicated its dream project in a media statement that went much deeper than the standard media releases for this glamorous market: "Napapijri's pioneering Circular Series of fully recyclable jackets has been recognized with the prestigious Cradle to Cradle Certified® certification gold level, the world's most advanced standard for safe, circular and responsible materials and products."[16]

"Developed and tested over the course of three years, Napapijri Circular Series is the first family of recyclable, mono-material products in the lifestyle fashion market to receive Cradle to Cradle Certified® gold from the Cradle to Cradle® Products Innovation Institute."[17]

The Napapijri Circular Series is unique because of its fully circular mono-material composition. The fillings and trim are made of Nylon 6 (Polyamide 6), and the fabric is made of ECONYL® Regenerated Nylon, a high-performance yarn that has been recycled from discarded fishing nets, landfill carpets, and other waste materials. Furthermore, as part of the ECONYL® Regeneration System. customers can return the jackets two years after purchase, so they can be transformed into new fabric through the ECONYL® Regeneration System.

---

[16] "Cradle to Cradle Certified® Certification", epeaswitzerland, Accessed June 4, 2024, https://www.epeaswitzerland.com/cradle-to-cradle-certified-certification/
[17] "Napapijri's recyclable jackets Cradle to Cradle certified", Innovation in Textiles, Accessed June 3, 2024, https://www.innovationintextiles.com/sustainable/napapijris-recyclable-jackets-cradle-to-cradle-certified/

Transforming Products – Safe and Circular Products - Case Studies

The Napapijri® brand story began in 1987, taking inspiration from the shadow of Europe's highest peak, the Monte Bianco. Originally a travel bag maker, they used design innovation mixed with style to revolutionize lifestyle apparel. Today, the Napapijri® brand is a variation of the Finnish word for the Arctic Circle, and the logo, half positive and half negative, gives visual expression to the North and South poles.[18]

Circular Series products, which started in Fall Winter 2019 as Napapijri's first circular fashion project, has since been recognized by the Cradle to Cradle Products Innovation Institute as one of the most innovative efforts at closing both product and technical loops. It is now set to become a fully-fledged family of circular products in upcoming seasons.[19]

Before this impressive accomplishment, no other brand has successfully developed a fully circular clothing system within such a short time span. Napapijri therefore succeeded in engaging, reviewing, and innovating across the technical cycle tiers.

Together with epeaswitzerland and a global network of forty-seven suppliers, Napapijri worked in only fourteen months to achieve certification from the Cradle to Cradle Products Innovation Institute. The suppliers came from four separate tiers, some of whom had to put aside competition in order to collaborate on the innovative and educational aspects of the project. Their collective effort resulted in the development of each component of the Circular Jackets, while strictly adhering to the Cradle to Cradle Certified® product standard, with epeaswitzerland developing the concept Supply Chain Domino Knowledge Transformation™ (See Chapter 12).

Starting with an in-depth assessment of the Circular Series' design and sourcing processes, the project also conducted an extensive mapping of

---

[18] "NAPAPIJRI", NRP srl brand & e-commerce images, Accessed June 3, 2024, https://www.rmpsrl.net/works/napapijri/#:~:text=The%20Napapijri%20brand%20began%20its,style%20and%20revolutionized%20outdoor%20apparel.

[19] "Napapijri Circular Series", epeaswitzerland, Accessed June 3, 2024, https://www.epeaswitzerland.com/project/napapijri-circular-series/

the Series' supply chain. All parties involved were committed to knowledge-sharing sessions, chemistry optimization, and joint development of product innovations to enable Napapijri to elevate the Circular Series to safer and better circularity standards. Napapijri decided in 2024 not to recertify the Circular Series Cradle to Cradle Certified®.

With over US$10 billion in turnover, VF Corporation is one of the world's largest apparel, footwear, and accessories companies connecting people to the lifestyles, activities, and experiences they cherish through a family of iconic outdoor, active, and workwear brands, Napapijri belongs to VF Corp.[20]

Roles of epeaswitzerland
Innovation Trustee

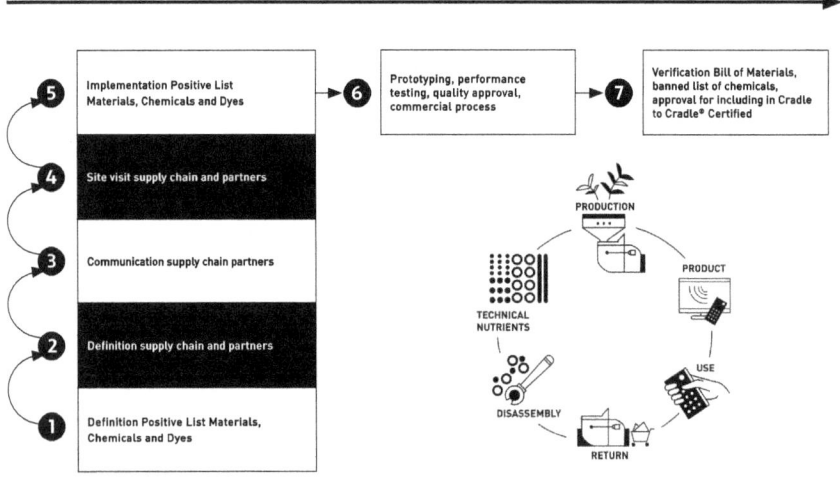

Figure 11: Innovation Process. Accelerating full speed in transforming timeframe of innovation projects down to 12 months.

---

[20] "Company", VF Corporation (VFC), Accessed June 4, 2024, https://www.vfc.com/our-company

Footwear

Essentials: Fighting against Goliath, Sport Brands

In the footwear market, the top ten brands account for more than 60 percent of sales. Nike is the clear leader in the industry, with a 22 percent share of the global market. Adidas is in second place, followed by Under Armour and Puma.[21] According to World Footwear's 2023 report, around 24 billion pairs of shoes were manufactured in 2022, over 60 percent in China, followed by 10 percent in Vietnam, 2.7 percent in Europe, and 1.4 percent in the US. China is the biggest consumer at about 18 percent, followed by 12 percent in the US.

During use, a pair of sneakers generates 125-175 grams of toxic microplastic through abrasion every year. The generated annual waste totals about 22 billion pairs of shoes. How can we get out of this Cradle to Grave dilemma? Many initiatives are trying to give these products a green touch, but most, if not all, are mere greenwashing.

How can we tackle the massive waste problem of chemical cocktails: the "foam problem," of polyurethane (PU), rubber soles, and glue?

A small Swiss company, Step Zero, wants to become a game changer. Step Zero aims to produce footwear components safe for health and circularity. Innovations will be required in materials and automated production processes with drastically reduced components. Footwear requires a complete redesign and innovation, which may influence how we live in the future. The products are defined to be safe for the biological cycle. Their plan to achieve this goal by 2026.

To succeed, Step Zero has yet to embark on their wildest adventure: converting their entire production to circular business. Every day, Step Zero works passionately to improve the circularity of products. The company has set the course and is ready for a circular future.

---

[21] "Global revenue of adidas, Nike and Puma from 2006 to 2022", Statista, Accessed June 4, 2024, https://www.statista.com/statistics/269599/net-sales-of-adidas-and-puma-worldwide/

Essentials: Walking on Cloud, Ski Boots

The project team cannot wait to see the world's first circular ski boot together with the inner liner on the retailer shelves. Using plastics repeatedly, the ski boot's outer shell is designed for the technical cycle, together with a ski-boot inner liner that is safe for the biological cycle.

In 2022, this was just a fun idea, and the stakeholders were naive enough to kick off the project in collaboration with Salomon, Heierling Ski Boots, epeaswitzerland, and the network of Next Generation. Step Zero developed the ski boot inner liner based on the Cradle to Cradle® Design Concept for Heierling and Salomon. After months of brain power, the product was tested on the slopes of the French Alps, surprising the professional skiers with an incredible new wear sensation. The materials like foam, soles, textiles, and accessories are designed to be safe for biological cycles. The product pre-launch is planned to take place in the winter season 2025.

Founded in 1885, Heierling is the oldest maker of ski boots worldwide. It all began in the winter resort of Davos in Switzerland and the company is still based there to this day. Heierling specializes in customized adjustments and fitting solutions for ski boots at the highest level. The company regularly serves some of the world's best ski racers. In fact, several Olympic gold medals have been won by racers wearing Heierling ski boots since the 1950s.[22]

Salomon SAS is a French sports equipment manufacturing company headquartered in Annecy, France. It was founded in 1947 by François Salomon in the heart of the French Alps and is a major brand in outdoor sports equipment. Salomon is owned by Finnish retail conglomerate Amer Sports, along with Wilson, Atomic, Precor, and others.

---

[22] "About us", Heierling, Accessed June 4, 2024, https://www.heierling.ch/en/company/about-us

Paper & Packaging

Essentials: Paper

The earliest known paper has been traced back to 200 BC in China. The craft of papermaking spread throughout the world and remained a relatively small-scale, artisan activity until paper production became industrialized during the 19th century. Today's top-quality paper requires a highly technical and accurate manufacturing process. New pulp, however, is an integral part of the paper making process: paper can be recycled only seven times at most because the fibers become shorter and deteriorate in quality. But even virgin and recycled paper are often toxic, especially when printed and coated. As one example out of many, thermal paper used for receipts contains Bisphenol A (BPA), which has effectively been banned in Europe since 2020. BPA has been a chemical of "special concern" for decades, although without scientific or regulatory consensus on its safety. BPA is one of the three bisphenols identified by the European Chemicals Agency (ECHA) as an endocrine disruptor.

Paper remains important but due to disruptive technologies and the digital age, it is under pressure. Global production of paper and cardboard amounts to more than 400 million metric tons per year.[23] Over 50 percent are produced in Asia, 24 percent in the US, 24 percent in Europe, and 1 percent each in Oceania and Africa. Packaging and board stock are the most common paper products. Due to the online shopping boom in recent years, demand is steadily growing.

According to World Wildlife Foundation (WWF), the pulp and paper industry consumes 40 percent of all wood harvested worldwide. This industry is also a major consumer of water, energy, and chemicals. It needs to reduce its environmental impact drastically, protect natural resources and the biosphere, and improve its economic performance. Innovative solutions are needed for water treatment and the use of biomass as fuel to advance decarbonization. Paper is a material of practical and sustainable choice and one of the most recycled. Recycling initiatives are widely

---

[23] "Global paper industry - statistics & facts", Statista, Accessed June 4, 2024, https://www.statista.com/topics/1701/paper-industry/#topicOverview

spread and achieved significant recycling yields of 68 percent in the US and 72 percent in Europe. However, hazardous chemicals in recycled paper and packaging endanger consumer health. Consequently, recycled paper is banned for direct food contact packaging.

How to transform such a giant industry and individual paper mills from a linear to a circular approach? Taking one paper product only and optimizing chemical inputs is a traditional, incremental innovation, but the largest impact and the real challenge would be to optimize all products used for manufacturing in multiple paper mills, in different locations, and even countries.

Which company would buy into an idea like this? To understand the challenge, one has to know that paper machines are monsters. A typical machine is about the length of two football pitches and around four meters wide. It can run speeds of up to 2,000 meters per minute.

The risk of failure with such an approach is high but, if successful, the result could be phenomenal. The global leader Mondi committed to the project with two paper mills in Austria and Slovakia. Implementing the Toolbox Cradle to Cradle™ approach of epeaswitzerland, 107 fresh and recycled products of the "Green Range" Paper Mondi got Cradle to Cradle Certified® at the bronze level. At the corporate level, Mondi defined the Mondi Action Plan 2030 (MAP2030). The proof of concept received an additional strong signal as two sites of Mondi South Africa were added later into the program, moving the project to a global scale.

Mondi is a global leader in packaging and paper, contributing to a better world. Mondi's 22,000 employees work across one hundred production sites in more than thirty countries, with key operations in Europe, North America, and Africa.[24]

---

[24] "Opportunity Unpacked: Mondi Group Integrated report and financial statements 2021", Mondi Group, Accessed June 3, 2024, https://www.mondigroup.com/globalassets/mondigroup.com/investors/results-reports-and-presentations/2021/integrated-report-and-financial-statements-2021/mondi-group-integrated-report-and-financial-statements-2021.pdf

Mondi's MAP2030 sustainability framework sets out the actions to meet ambitious sustainability goals by 2030. This includes 100 percent packaging and paper products that are reusable, recyclable, or compostable, integrating circular-driven solutions of innovative packaging and paper to keep materials in circulation and avoid waste.

Mondi is also committed to creating an empowered and inclusive team that contributes to a better world. The company wants to inspire a global and diverse workforce, which feels safe to develop the skills they need now and in the future. Mondi supports people in realizing their individual potentials and aims to create an inclusive work environment that provides equal opportunities and values safety, health, and mental wellbeing.

The company is committed to protecting its environment to mitigate climate change and reduce its ecological footprint.

Mondi acknowledges that, as a business, they play a critical role in addressing climate change, setting targets, and innovating to reduce greenhouse gas (GHG) emissions and help drive the transition to a low-carbon economy. Taking committed steps to reduce CO_2 emissions will help minimize Mondi's contribution to climate change and improve efficiency to reduce energy use and deliver value through sustainable products and solutions. Maintaining ecosystems' resilience is critical to mitigate climate change and to achieve Mondi's climate goals. [25]

The focus lies in innovation to develop new solutions, keep materials in circulation, and engage with customers and partners to drive progress at scale. Mondi considers the environmental impacts of products at each stage of the value chain, from the sourcing of raw materials to material efficiency, product design and safety, and a sustainable end-of-life.

Essentials: Toilet Paper – Chemicals in Water

Oceans and other bodies of water being polluted with chemicals, pharmaceuticals, microplastics, and trash, are a growing concern. The

---

[25] "MAP2030 framework", Mondi Group, Accessed June 3, 2024, https://www.mondigroup.com/sustainability/map2030-framework/

impact is severe for society and the planet. In an *Environmental Science & Technology Letters* publication from March 2023, results were presented showing that toilet paper is a potential major source of PFAS (per- and poly-fluoroalkyl substances), also known as "forever chemicals," entering wastewater treatment systems. This is bad design from the start. Transformation of the hygienic paper industry is required. A positive light is the hygienic paper mill, Lucart, located in France, where the Toolbox Cradle to Cradle™ concept was applied for the entire manufacturing plant, followed by a Cradle to Cradle Certified® certification bronze level aiming for silver level for the recertification. This was the first big step towards a safe and circular future. Given the initial gains, a more in-depth project was approved, which will be described later in a case study about beverage packaging recycling.

Lucart, a multinational group promoting sustainable paper making, was founded by the Pasquini family in 1953. The company takes a systemic approach designed to optimize the use of raw materials and reduce waste by turning it into a new secondary resource.[26] Lucart makes air-laid and MG (machine glazed) paper tissue products (paper articles for everyday use, including toilet paper, kitchen paper, napkins, tablecloths, handkerchiefs, etc.).[27] The production activities are organized across three business units (Business to Business, Away from Home, and Consumer), engaged in the development and sale of products with brands like Tenderly, Tutto, Grazie, Natural and Smile (Consumer market) and Lucart Professional, Fato, Tenderly Professional, and Velo (Away from Home market). The paper production capacity is over 396,000 tons/year on twelve paper machines. Seventeen hundred people are employed at ten production plants (five in Italy, one in France, one in Hungary, two in Spain, and one in the UK) and a logistics

---

[26] "Intesa Sanpaolo and Sace Support Sustainable Development of Lucart", Intesa Sanpaolo, Accessed June 4, 2024, https://group.intesasanpaolo.com/content/dam/portalgroup/repository-documenti/newsroom/comunicati-stampa-en/2022/1/CS%20Finanziamento%20ISP%20e%20SACE%20a%20Lucart%20(EN).pdf

[27] "Lucart chooses Enel X as it continues on its decarbonization path", Enel, Accessed June 4, 2024, https://www.enel.com/es/medios/explora/busqueda-comunicados-de-prensa/press/2022/01/lucart-chooses-enel-x-as-it-continues-on-its-decarbonization-path

hub. The consolidated turnover is over €550 million. Lucart has been awarded a "Platinum" sustainability rating by the independent certification company Ecovadis.

Essentials: Offset Printed Paper for Food Packaging

Do you like pizza? Please watch out for the food packaging, even if it's made with virgin paper, since it might contain toxic chemicals caused through paper contaminants and migration of printing inks. Authorities have banned using recycled paper for packaging with direct food contact but are granting approvals for virgin packaging as food grade.

Who can break out of this endless wheel and find a feasible industrial solution? The tiny offset printing company Voegeli, based in Switzerland, successfully tackled the problem. Voegeli AG has always played a pioneering role when it comes to sustainable offset print products. The family-owned company with around fifty employees, based in Langnau Emmental, Switzerland, was founded in 1911. Voegeli AG produces offset printed products such as brochures, magazines, annual reports, and packaging.

Voegeli has been printing and producing without using the VOC (volatile organic compounds) substance of isopropyl alcohol and has installed solar panels and heats its building fossil-free, uses groundwater for cooling, and has been operating "with no fossil fuels" for over twenty years.

At that time, Voegeli began to "rethink" their products to reach the next level in sustainability. What followed was nothing less than the most important change in the company by far: the decision to completely convert production to Cradle to Cradle®–the holistic approach to safe and circular products without any loss of value.

Inks, paper and cardboard, varnishes, and adhesives are optimized to be safe for human beings, the environment, and biological cycles. In 2019, Voegeli AG's top world ranking earned the company the Cradle to Cradle Certified® gold level for offset printing and packaging products. The printed soft- and hardcover book has been printed by Voegeli AG and is Cradle to Cradle Certified® at gold level.

Essentials: Beverages Packaging Paper, Plastics, Aluminum – Fighting against Goliath

Proper packaging is crucial for the delivery of innovative, safe products to the consumer. Safe food packaging needs to fulfill three functions: protection, preservation, and promotion of the product. Intelligent performance packaging for beverages and food packaging can prevent product deterioration caused by microbiological growth, chemical contamination, protect food from light, moisture, leakage, and physical damage, while preserving durability of products. Other factors that must be considered are economics, storage, and logistics. In addition, regulations preventing paper board packaging coming into direct contact with food must be followed.

Three global players—Tetra, SIG, ELOPAK—are in the beverage packaging paper business and have a global market share of 60 percent. To remain successful in the business requires high tech, financial investments, and integrated business models.

One particular innovation changed consumers' daily lives: a multilayer composite consisting of paper, aluminum, and plastic. The concept is more complicated than just three layers. For example, a possible layering sequences would be polyethylene, paperboard, polyethylene, aluminum foil, polyethylene, and doubled polyethylene.

Are these products compatible with material health and circularity? They are approved for food safety but do not meet Cradle to Cradle® design standards. A redesign is required, and from our perspective, no perfect solution is yet available

Nonetheless, effective business recoups resources through collecting used packaging materials from consumers and inventing new products through recycling. Lucart, the hygienic paper mill in France, has developed product solutions from beverage cartons.

"EcoNatural" is the paper and dispenser system that gives new life to beverage cartons. Thanks to an innovative technology, Lucart separates the components of beverage cartons. From cellulose fibers, it creates Fiberpack®, which it uses to produce paper, and from aluminum and polyethylene it produces Al.Pe.®, the new raw material used for making dispensers.

Plastics

Essentials: Eco Junk – Flexible Packaging, Multi Material Layers, No Circularity, Loss of Resources

Flexible packaging allows end-users to bend, fold, and shape without breakage or other kinds of damage. Flexible packaging saves resources compared to other kinds of packaging. However, flexible packaging laminates are a "no go" for Cradle to Cradle®. In the late 1950s, Procter & Gamble first designed multilayer tubes for toothpaste. Flexible packaging solutions are multi-layer combinations of new materials, composites, and hazardous materials, using innovative technologies to give barrier properties, strength, and storage stability for food or other consumer goods. Most multi-layer flexible packaging has been created without considering recyclability. Consequently, 2.6 million tons of multilayer film waste are incinerated or landfilled every year.

Chemicals in current plastics production are a critical issue. Between five thousand to six thousand chemical substances are used to produce plastic products. Among them are many health-related substances. This does not include the metabolites (breakdown products) caused by the direct contact of plastics with the environment (for example, water, UV radiation) or upon heating (for example, dyeing in the textile industry or in the microwave oven).

So far, little attention has been given to this issue, and legislation appears to support the status quo. The industry is under no pressure to make any changes. For a plastic product's twenty most important functional characteristics, as many as five hundred chemical substances whose safety can be guaranteed in the biological system would be needed.

Essentials: The Breakthrough, Making the Impossible Possible, Mono Materials, Printing Inks, Selecting Technology

Achieving another milestone on its journey to a genuine circular economy, Werner & Mertz has created the first flexible packaging in the world to receive a gold rating in all categories of the Cradle to Cradle Certified® product scorecard. In line with the Recyclate Initiative, the fully recyclable pouch is made of a mono-material (polyethylene) with a removable wrap-around label. The pouch is the result of a five-year joint project by Werner & Mertz and the global packaging and paper company, Mondi. The innovative, patented pouch design solves the problem of recycling printed plastic by reducing the printed area to only 15 percent and removing adhesives and bonding agents from the printed label. As a result, 85 percent of all packaging material can be mechanically recycled without any loss of quality.[28]

The independent Institute Interseroh and HTP Cyclos confirmed the pouch's complete recyclability. Interseroh certified that the stand-up pouch has "excellent recyclability that cannot be optimized any further."[29] I described the stand-up pouch as an actual "lighthouse project."

Reduce-Reuse-Recycle

The patented stand-up pouch is the first packaging that complies with the three Rs: reduce, reuse, and recycle. Compared to a bottle, the

---

[28] "About Us," Werner & Mertz, accessed June 3, 2024, https://werner-mertz.de/en/about-us/
[29] "Interclean 2020: Amsterdam Innovation Award", Werner & Mertz Professional, Accessed June 3, 2024, https://wmprof.com/se/interclean-2020-amsterdam-innovation-award/

pouch saves 70 percent on packaging material for the same amount of product (reduced). A product bottle can be refilled repeatedly thanks to the refill packaging (reuse). Lastly, even the plastic of the pouch remains in the closed material loop (recycle). And yet, there is another unique feature: the original bottle for Werner & Mertz Frosch's consumer brand is 100 percent recycled plastic from post-consumer household waste collections.

Award-winning Innovation

The first mono-material refill pouches for Frosch brand products have been on retail shelves since November 2019. The innovation has already received several notable awards.

The German Packaging Award in 2019 said: "The great challenges of our time are packaging developments perfectly adapted to the recycling economy. This bag impressively demonstrates what can be achieved today with a consistent recycling approach to design." The pouch won in the Sustainability category.

At the Interpack trade fair in Düsseldorf, the pouch won the international World Packaging Award. "The highly innovative 100 percent recyclable pouch has cleared sorting and recycling hurdles on its way to genuine recyclability…That is Cradle to Cradle®!" said the competition's World Packaging Organization.

The German Design Awards 2021 gave the prize to the Frosch brand in the category "Excellent Communications Design – Eco Design"! The manufacturer Werner & Mertz of Mainz has received the award for "outstanding design quality" from an international jury of experts.

The pouch concept revolutionizes the recycling industry

The design process was reverse engineering: starting from the end of the product life cycle, then creating a packaging fit for every stage of the recycling process. Every stakeholder along the entire supply chain was involved, making it possible to design packaging suited to recycling and the circular economy. The development of the stand-up pouch even influenced the design of sorting facilities. Previously, sorting and recycling equipment was unable to process plastic sheeting smaller than letter-sized paper; instead, the material was incinerated for recycling. "Plastic sheeting makes up about 40 percent of the waste in the Yellow Bag (Germany) and represents enormous but untapped potential!" explained Immo Sander, former developer of packaging at Werner & Mertz. "In the sense of Design for Recycling, it was important to us to develop a stand-up pouch that was not just theoretically recyclable. It also had to ensure practical implementation of the recycling process." During the years of development work, designers consulted with manufacturers of sorting systems to guarantee separate sorting and recycling.[30]

In 2020, Werner & Mertz began making all its Frosch brand pouches in the new design, even expanding the conversion to include its green care brand for professional use. The company has no intention of resting on its laurels but is working on more extensive plans. In the future, it intends to produce flexible plastic packaging that will not only be completely recyclable but also made of recycled materials to begin with. When enough packaging goes into recycling systems that comply with Design for Recycling guidelines, this packaging solution can be realized with safe recyclates, eliminating the need for new materials. "In this chicken or egg situation, we lay the foundation with our pouch and hope for imitators," says said Werner & Mertz owner Reinhard Schneider "After all, as far as climate change is concerned, we have no choice but to increase recycling rates significantly."

Development History of the Patented Stand-up Pouch

The first completely recyclable stand-up pouch of a monomaterial (polyethylene) with a removable wrap-around label (also of polyethylene),

---

[30] "Interclean 2020: Amsterdam Innovation Award", Werner & Mertz Professional, Accessed June 3, 2024, https://wmprof.com/se/interclean-2020-amsterdam-innovation-award/

designed according to the Cradle-to-Cradle® principle, was a joint development of Werner & Mertz and the packaging and paper company Mondi.

"With this groundbreaking design for recycling, we have taken a giant step toward closed-loop circulation of plastic packaging," said Sander. After use, the patented stand-up pouch can be recycled 100 percent into recyclates that have almost the same quality as the raw material.

This innovation brings to reality an idea that has long been on the Werner & Mertz agenda. "Back in 2014, we at Werner & Mertz were looking for 'single-material concepts'," said Sander. The goal was a monomaterial pouch, instead of the usual multi-layer product, that could be printed and filled using existing technologies. It quickly became apparent that the goal could be achieved only in cooperation with a packaging specialist. Werner & Mertz found that in Mondi. "Nowadays, it is hardly possible for just one company to develop such an innovation independently," notes Thomas Kahl, Project Manager of EcoSolutions from Mondi Consumer Packaging. "It takes a network to develop packaging for the circular economy. And Mondi and Werner & Mertz were prepared to take up the challenge."

In 2015, the two companies established a project group with The Green Dot, epeaswitzerland, and cyclos-HTP. The three partners supported the development with advice on material selection and confirmation of the pouch's recyclability and integration into existing recycling structures. The first thing the project participants did was to find out precisely what "Design for Recycling" meant. Even the experienced packaging experts had to learn to think the product through from the other end. After an intense development phase in 2017, a stable and entirely usable stand-up pouch made of polyethylene was created.

After overcoming all bureaucratic hurdles, Mondi and Werner & Mertz presented the new product to the public in 2018. The first pouches were made available in retail markets in autumn 2019.[31]

---

[31] "Patented Stand-up Pouch: The ecological all-rounder", Werner & Mertz, Accessed June 3, 2024, https://werner-mertz.de/en/patented-stand-up-pouch/

Cleaning Detergents

Essentials: Cleaning Detergents, Safe for Biological Cycles, Degradable in Water

Cleaning detergents are substances or a mixture containing soaps and/or organic substances intended for washing and cleaning. Some ingredients commonly found in many cleaners can cause respiratory irritation and contribute to water pollution. One alternative is phosphates with low toxicity, but they cause nutrient pollution and feed algae, which leads to eutrophication and harmful algal bloom (HAB). The EU and US have banned the use of phosphates in detergents.

In 1986, Werner & Mertz introduced the first phosphate-free household cleaning agent under the brand name Frosch.[32] They have always used natural ingredients such as vinegar, lemon, and soda (natron) in their products.

Many cleaning products consist of substances called "surfactants." Surfactants cause grease and grime to dissolve in water during washing or cleaning. Conventional cleaning agents often contain surfactants made from petroleum. Sustainable cleaners, however, are made from regenerative raw materials—i.e., renewable plant-based sources. In addition to using petroleum, cleaning product manufacturers have produced surfactants from tropical palm kernel oil and, to a lesser extent, coconut oil. However, the conventional, non-sustainable cultivation of tropical oils destroys the rainforest's biodiversity in the long term and even the rainforest itself. For the sake of biodiversity, Werner & Mertz continually increase the ratio of surfactants based on European oil plants such as rapeseed, olive, linseed, and sunflowers. They also use sustainable tropical oils. None of these oils competes with food production.

Sustainability is often equated with efforts to reduce environmental damage. However, action must be taken much earlier to keep harmful substances from circulating in the first place. A genuine circular economy

---

[32] "Environmental responsibility", Werner & Mertz - Wikipedia, Accessed June 4, 2024, https://en.wikipedia.org/wiki?curid=21670797

means doing the right things in the right way from the start. Werner & Mertz lives this credo throughout the entire value-added process, from raw materials to packaging, in accordance with the Cradle to Cradle® principles. It begins with the use of renewable energy in the production process and efficient in-house water treatment and encompasses the sustainable ingredients in the formulas, and packaging made from recyclate. Moreover, Werner & Mertz commits to wide-ranging social and ecological justice and to conserving biodiversity. The company is convinced that the circular economy integrating the C2C® principle is the most sustainable way of doing business.

Since 2013, Werner & Mertz has had its Frosch and Green Care Professional Brand stamped Cradle to Cradle Certified®. It was the first company in the cleaning industry in Europe to receive Cradle to Cradle Certified® Gold for the Frosch Citrus Shower & Bath Cleaner. In the same year, eleven Green Care Professional products also received gold certification. In 2024, over sixty products bear this highly coveted distinction.[33]

Essentials: The Fragrance Story

In high-quality cleaning detergent products, fragrances are essential. They are present in small quantities and need to have a low allergenic risk to be safe for human beings. Some products do not contain any fragrances at all. Today, consumers required sustainability and complete transparency, declaration of origin of agriculture products, impact on health and environment. But the fragrances industry has a "top secret" culture, and confidentiality is the backbone of their business. How can a company like Werner & Mertz be assured that their marketing statements on their products are accurate if they do not know what "confidential substances" are in the fragrances? The only way to find out is through testing.

To get a Cradle to Cradle Certified® certification process for cleaning detergent products from the brand Frosch, the fragrances formulas needed to be disclosed. Not an easy task. How to get the yes? After

---

[33] "In Perfect Harmony with Nature", Circular Economy - Werner Mertz Professional, Accessed June 4, 2024, https://wmprof.com/sustainability/circular-economy/

extensive discussions with various companies, the knowledge trustee role of epeaswitzerland as independent and impartial third party led to this information being supplied. The result allowed the Cradle to Cradle Certified® certification at gold level. The Frosch marketing team was reassured that the science was reliable, marking another milestone for epeaswitzerland's credibility and trust in an industry business environment.

Wood

Essentials: Parquet Flooring – 100 Years to Grow, Four Generation Lifetime to Use

According to Forest Research (www.forestresearch.gov.uk), the carbon stocks of forest biomass have increased in Europe, North and Central America, and Asia between 1990 and 2020. Yet, on a global level, stocks decreased over this same period by around 4.7 million hectares (0.1 percent) per year. Of the 3.9 billion cubic meters removed from global forests, 49 percent were used as fuel, and the rest by industrial roundwood processors. Global wood production amounted to 473 million cubic meters of sawn wood (25 percent from Europe) 368 million wood-based panels (22 percent from Europe) and 401 million tons (21 percent from Europe) went to the paper and paperboard industry. At the going rate of wood consumption, resources will become even more precious and costly.[34]

Bauwerk Group, a leading producer and supplier of premium wooden flooring, has manufactured parquet flooring for generations. Bauwerk Group employs more than 1,750 people in its many divisions all over the world. The Group, headquartered in St Margrethen, Switzerland, includes the Bauwerk Parkett and BOEN brands, and since May 2022, the North American company Somerset Hardwood Flooring.

Selling some 9 million square meters of wooden flooring every year, the Group offers a comprehensive product range with solid two-layer and

---

[34] "9.4 Forest carbon stocks", Forest Research, Accessed June 4, 2024, https://cdn.forestresearch.gov.uk/2022/12/FS2022-combined-29sep22.pdf

three-layer parquets, as well as specialized hardwood floors for sports facilities. It has production plants in Switzerland, Lithuania, Croatia, and the US. With the addition of Somerset Hardwood Flooring, Bauwerk Group has reached a turnover of 350 million Swiss francs, solidifying its position as the market leader in quality parquets.

The company has developed a flooring solution that not only forgoes any toxic substances in its products, but also does not produce waste to be incinerated. This technology allows the flooring to be dismantled, processed, and reused at the end of its service life. The same parquet flooring can be reused and refurbished across at least three to four product generations, each twenty-five years, without any loss of quality. Considering it takes one hundred years for an oak tree to grow, this product innovation allows a doubling of the lifetime of wood resources. Furthermore, through a process known as cascading, the material of the parquet can be reutilized for other products. The parquet is designed so the materials always stay within a closed loop. No individual parts end up being discarded, unnecessary energy usage is prevented, and the CO_2 contained in the wood remains stored over decades and even centuries. Functionally, the parquet flooring absorbs the sound and gives comfort. Bauwerk Parkett developed its new technology in conjunction with epeaswitzerland in line with the Cradle to Cradle® Design Innovation concept. For two of its products, Cleverpark Silente and Multipark Silente, Bauwerk Parkett was the first wooden flooring manufacturer to receive the Cradle to Cradle Certified® gold certificate. Other products in the Bauwerk assortment are available at silver and bronze level. The certification process also involved an analysis of Bauwerk Group's suppliers, enabling a transparent supply chain and a material health assessment of all chemical ingredients used in the product.

Essentials: Wooden Chipboard – Ever heard of Formaldehyde or Composites?

Formaldehyde is a substance that naturally occurs in wood. Low doses of this substance are harmless, but higher concentrations of formaldehyde are cancerogenic. In the panel manufacturing industry of

Multi-Fiberboards (MDF) and Low-Density-Fiberboards, also referred to as chipboards or particleboards, formaldehyde is used for binders and glues. Europe, the US, Japan, and Australia are setting and supervising emission standards to reduce formaldehyde in products.

Melamine-Faced Chipboard is an important material in the furniture and interior design industries. Made from particleboard that's been covered with decorative paper and impregnated with melamine resin, it is the most widely used material for modern furniture, especially because it is inexpensive. Melamine resin is a heat-resistant hardening resin formed by the condensation of melamine and formaldehyde.[35]

In 2021, European countries officially banned cups and plates that contain melamine and bamboo. This is because, if such containers are used for a long time to hold hot food, or if their surfaces are scratched and worn, traceable amounts of melamine and formaldehyde will be released into the food.

Wooden chipboards are cheap and manufactured as mass products, widely mixed with materials like wood, which belong to the biological cycle, and other materials such as resins and laminates belong to the technical cycle. It is a real challenge to innovate wooden chipboards and laminates to be safe and circular!

Innovations are transforming mass production, with biobased and biodegradable materials that meet performance requirements and are affordable. Just putting Cradle to Cradle® material solutions together, one by one, could work on a theoretical level. But this process will also require additional equipment for mixing and drying.

Let's design a Cradle to Cradle® wooden chipboard: Use FSC-certified wooden chips combined with proprietary biobased and biodegradable resins, which could be used as laminates as well and printed decorated paper. And how about closing the loop with furniture or kitchen surfaces, where the lamination is added by local decentralized craftsmen, sold to

---

[35] https://www.compositepanel.org/products/particleboard/

a selective market or by large furniture manufacturers in outlet furniture stores? How should a thorough take back system be organized? How should the materials be reintegrated back into the supply chain or return to biological cycles? A confidential project is now underway. Hopefully, the end-product can be implemented and become a lighthouse at an affordable price in a global competitive mass market.

Essentials: Wooden Ceilings – Reintegration Back into the Supply Chain

Ecology and environmental protection have always been an important element in the business philosophy of Knauf Ceiling Solutions. Headquartered in Germany, the company crafts innovative ceiling solutions across eleven state-of-the-art facilities in Europe and Asia. With over eighteen hundred dedicated professionals, Knauf Ceiling Solutions is a multi-material provider offering a harmonized portfolio of proven product brands into four categories: Mineral Solutions, Metal Solutions, Wood and Wood Wool Solutions, and System Solutions.

Knauf Ceiling Solutions provides ecological and environmental-friendly products by adopting resource-safeguarding manufacturing processes and utilizing natural raw materials like magnesite, water, and wood-wool. Wood-wool and magnesite imbue decorative panels with many favorable features: flame retardant, moisture absorption, sound absorption, climate regulation, and material stability. Knauf Ceiling Solutions products branded Heradesign® are manufactured in Austria. They comprise an innovative ceiling system that not only provides a rich design variety but also have excellent sound absorption effects. These products are thus favored by architects, planners, and construction workers. Heradesign® systems are used for building segments that require particular robustness, such as attics, halls, corridors, and gymnasiums.[36]

---

[36] "Heradesign Wood-wool Panels", Knauf, Accessed June 4, 2024, https://www.knauf.com.cn/en/product/show.aspx?cid=72339069014638592&cid2=72341268037894144&id=85

As for material health and circularity, some of the colors used in Heradesign® products are Cradle to Cradle Certified® gold level. With a special process technology installed on the Austrian premises of Knauf Ceiling Solutions, the post-industrial or post-consumer waste produced by magnesite binder and wood wool are used to generate renewable energy, with the magnesite binder reintegrated back into production without any loss of quality. Take-back solutions from construction buildings are still at an early stage, but will expand in the future, driven by ambitious legislation frameworks.

Construction Industry, Building Materials

Essentials: Cement - A Dead-End Road... Monitoring not Legal Limits, Ashes from Toxic Waste as Filler for Cement

The cement industry is one of the largest global industrial polluters, responsible for between 4 to 8 percent of worldwide man-made carbon dioxide (CO_2) emissions. About half stems from the chemical process and at least 40 percent from burning fuels. If the cement industry were a country, it would rank third as a CO_2 emitter in the world—up to 2.8 billion tons, placing it after China and the US.

To fight this emissions dilemma, engineers have explored substituting materials through fly ashes, which are waste products left over from burning coal, generating energy through the incineration of waste, like rubber tires and other kinds of waste. These ashes are used as cheap filling materials, which have similar qualities as cement, but which also contain toxic heavy metals. What about the toxicity of the heavy metals, which might be leaching out of buildings, into the living environment, or through infrastructure into the environment?

Incineration plants are burning municipality and/or industrial waste to secure a level of protection, especially from toxic and persistent organic pollutants such as mercury and polychlorinated dioxins and furans. Incineration achieves this through Best Available Technologies (BAT) including filtration of waste particles at pre-defined legal limits. Still, this process sends emissions into the environment. The cement industry negotiated agreements with authorities to monitor and report emissions from cement manufacturing, but with no requirement to adhere to legal limits.

Is there any hope for sustainable cement production and materials? The Global Cement and Concrete Association (GCCA), headquartered in London, was founded in 2019 to be the global voice of the cement and concrete sector. One of the Association's objectives is to develop and strengthen its contribution to sustainable construction across the value chain. The GCCA aims to foster innovation throughout the construction value chain in collaboration with industry associations, as well as architects, engineers, developers, and contractors.

Could Cradle to Cradle® provide a solution for safe and circular cement, manufactured with renewable energy and no burning of waste or fossil fuel? Yes, but the transformation will be radical, fundamental, and needed for future generations.

Ceramics

Essentials: System Integration into a Global Industrial Group

Ceramics is one of the most ancient industries, going back thousands of years when humans discovered that clay could be formed into objects by mixing with water and heating with fire. The oldest known ceramic artifact dates to 28,000 BCE. Since those prehistoric times, the ceramic industry has become an integral part of technology, art, and industry. The Industrial Revolution was the turning point when mass ceramic production, together with advanced kilns and glazing applications, started. Today, ceramics manufacturers are still primarily small and medium-sized enterprises (SME). Ceramics are found in a wide range of products: medical devices, high tech electronics, kitchen and sanitary ware, tiles, bricks, clay pipes, and porcelain.

Growing environmental concerns about raw materials extraction, intensive use of energy, water, and greenhouse gas emissions are forcing the ceramic industry to take huge steps towards a sustainable future. Initiatives to reduce the environmental impact are growing and recycling technologies are being developed.

Being part of an emission intensive construction industry, the well-known brand LAUFEN, which is part of Roca Group, is committed to making a significant contribution to a more sustainable world. Roca Group started in 1917, manufacturing cast iron radiators for domestic heating in its factory in Gavà, Barcelona. In 1929, it entered the bathroom sector with the manufacturing of tubs, ceramics (1936), and water faucets (1954).

In the 1990s, the Group carried out its first international expansion phase, mainly based on the opening of subsidiaries and acquisitions of leading companies in their markets.

In addition to Portugal and France, the company started business in the United Kingdom, Germany, Italy, Morocco, Argentina, Brazil, and China. The turning point in the internationalization process came in 1999 with the acquisition of the Swiss group, Laufen Keramik Holding, the world's fourth largest manufacturer of ceramic sanitary ware at that time. The purchase of LAUFEN allowed Roca Group to consolidate its presence in strategic markets where it was less established like in Eastern Europe, Brazil, and the United States.

In 2005, Roca announced an ambitious strategic plan to focus on the bathroom sector, progressively divesting all other businesses, and in 2006, they achieved world leadership in the sector. Today, Roca's commercial network reaches more than 170 countries, supplied by its seventy-nine production plants and its 20,000 employees all around the world.

LAUFEN's sustainability strategy focuses on innovations that enable the company to efficiently transition towards a circular economy and comply with the UN Sustainable Development Goals (SDGs) while pursuing growth to enable long-term success. To magnify the impact of its own efforts, LAUFEN is joining forces with its parent company, Roca Group.

Within a circular economy project based on Cradle to Cradle® Design principles, LAUFEN has chosen a pilot manufacturing plant to implement the radical transformation, focusing on material health, product circularity (including the integration of post-consumer materials back into the supply chain), clean air and climate protection, water and soil stewardship, and social fairness.

The ambitious goal is to create proof of concept and use it as blueprint to for other production sites within Roca Group to enable a safe and circular future.

Nowadays Roca Group continues to be a 100 percent family-owned company engaged in the creation of bathroom spaces, an activity that has made it a global leader.[37]

Essentials: Facades – An Industry Responsible for a Lot of Waste, Take Back, Reintegration into Supply Chain

Within the construction industry, sustainability has become more focused for both giant projects and single-family homes. Step by step renewable energy, water stewardship, materials and circularity are becoming a highly needed push. The entire construction industry represents about half of all raw materials extracted from the earth's crust. Construction and demolition activities represent 50 percent of all waste generated, 35 percent of energy consumption, and more than 50 percent of global carbon emissions. This is a dead-end. How can this dilemma be taken and turn it around to create a positive impact?

Trimo, a manufacturer of prefabricated modular sandwich elements in Central Europe, nestled between the Alps and the Adriatic, has started to take action with a pilot project for safe and circular building materials according to Cradle to Cradle® design principles, integrating a take back approach to reintegrate the materials into the supply chain. Established in 1961, Trimo is one of Europe's leading companies

---

[37] "About Roca", Roca Accessed June 3, 2024, https://www.roca.co.id/aboutroca

offering innovative, highly efficient, sustainable envelope systems (wall and roof) and modular space solutions.

More than 60 million m² of facades and roofs have been produced over the past 63 years in over 120 countries worldwide to more than 25 thousand customers and partners. Trimo has established a sales network in almost thirty countries and has production facilities with around 450 employees in Slovenia and Serbia.

Electronics

Essentials: Scare Resources, Mission Impossible, Global Supply Chain Transformation – Rethinking the Way We Make Things

Electronics is now one of the largest industries in the global economy. As more societies become digitally and smartly connected, the boom will continue. Electronics are generating more revenue than any other manufacturing sector, and an estimated eighteen million people are employed in the sector. However, consumer electronic brands are facing many supply chain challenges in a global competitive and prospering market. Original Equipment Manufacturers (OEM), a product manufacturer for leading brands, engages hundreds of suppliers within the supply chain, who must deliver on a just-in-time system. A similar phenomenon occurred in the fashion industry, resulting in supply chain caravans moving to cheaper production locations.

One consequence of such developments has been a massive toxic disaster.

According to the UN, eight kilograms of e-waste per person was produced worldwide in 2023. This means 61.3 million tons of electronic waste are discarded every year. Only 17.4 percent of this waste, containing a mixture of harmful substances and precious materials, is properly collected, treated, and recycled globally.[38] The remaining 50.6 million tons will be placed in landfills, burned, illegally traded, treated in a sub-standard process, or simply hoarded at home. Even in Europe, which leads the world in e-waste recycling, only 54 percent of e-waste is officially reported as collected and recycled. A lack of public awareness is a major barrier preventing countries from developing circular economies for electronic equipment.[39]

Is there a light in the tunnel? In 2012, SENS eRecycling, a foundation that organizes the recycling of household appliances in Switzerland, started a search field project with epeaswitzerland. How could the electronic industry, including the electronic recycling and reintegration of materials, into the supply chain be transformed? The results, obtained over a decade ago, are still valid today, since not much changed over the years. The SENS Foundation, based in Switzerland, had expertise in established take back systems from retailers and close contacts with the electronic recycling industry. epeaswitzerland's expertise was in enabling transparency in the supply chain, together with knowledge in circularity, business models and Cradle to Cradle®. The feasibility study included the circular accounting by epeaswitzerland™ methodology (see Chapter 12), which could prove the economic benefits of electronic goods in a circular context.

The SENS Foundation plays a critical role in setting pioneering benchmarks in eRecycling. As an expert in the sustainable recycling of post-consumer electrical and electronic appliances as well as lamps, lighting equipment and photovoltaic systems, it conserves resources and thus makes an important contribution to environmental protection.[40]

---

[38] "International E-Waste Day to shed light on 'invisible' electronic waste", Recycling Magazine, Accessed June 4, 2024, https://www.recycling-magazine.com/2023/06/27/international-e-waste-day-to-shed-light-on-invisible-electronic-waste/

[39] "Invisible e-waste unveiled - International E-Waste Day proves success once again!", weeeforum, Accessed June 4, 2024, https://weee-forum.org/ws_news/invisible-e-waste-unveiled-international-e-waste-day-proves-success-once-again/

[40] "Welcome to SENS eRecycling", SENS eRecycling, Accessed June 3, 2024, https://www.erecycling.ch/en/

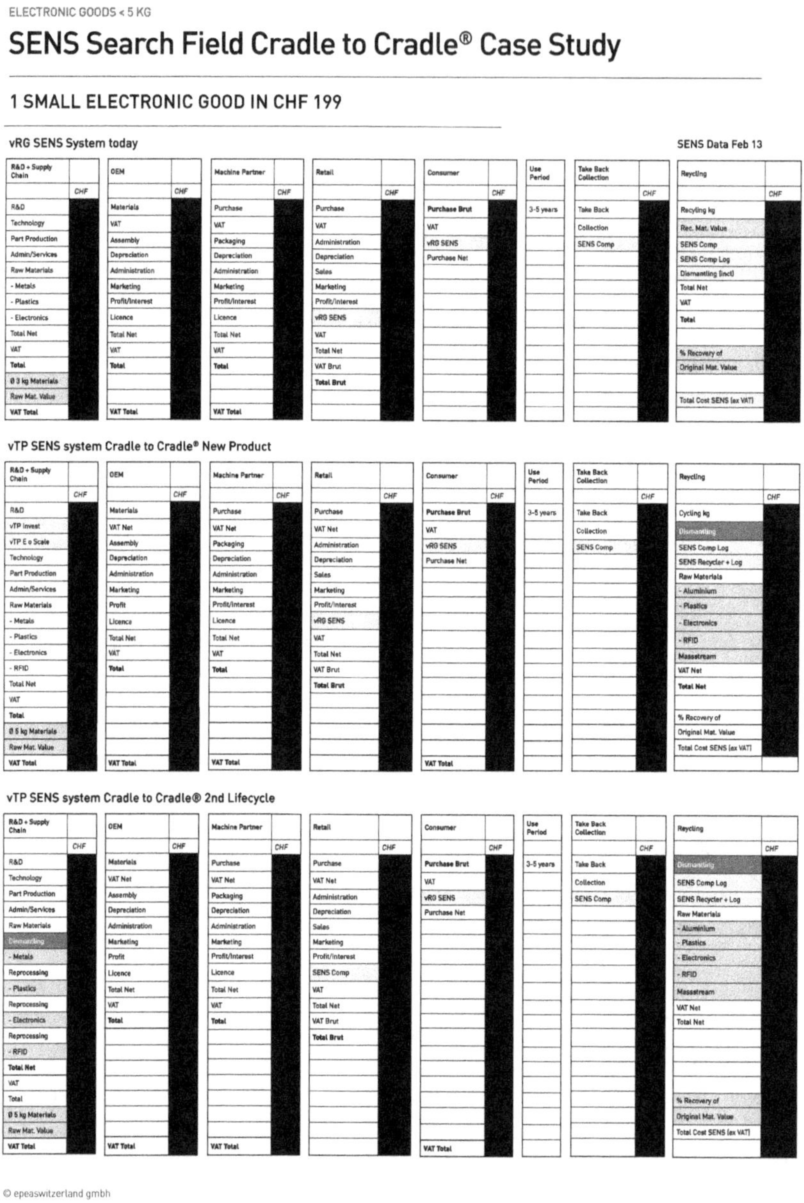

From Rebel to Radical Innovator

ELECTRONIC GOODS < 5 KG

SENS Search Field Cradle to Cradle® Case Study

1 SMALL ELECTRONIC GOOD IN CHF 199

Figure 12: Circularity with Circular Accounting by epeaswitzerland™.

The WEEE Forum (Forum for Waste Electrical and Electronic Equipment) is a European association of forty-one systems for collecting and recycling electrical and electronic waste. Its task is to provide and maintain a center of excellence to promote cooperation and dialogue on tried-and-tested practices. The SENS Foundation is a founding member of the WEEE Forum and presided over it for four years starting 2012.

Linear accounting proves unfit for circular projects. epeaswitzerland developed Circular Accounting by epeaswitzerland™ for circular economy.

When the results of the study were presented, I strongly expected that other electronic projects would follow. However, ten years passed before the first company came on board. I experienced a rollercoaster of emotions, including hope and disappointment, but with the conviction that more companies will come prevailed.

Flooring

Essentials: Parquet Flooring – Lacquer Natural versus High Tech Chemistry

The world of lacquer natural ingredients versus high-tech chemistry provides an unexpected challenge for circularity.

Cleverpark Silente and Multipark Silente products of Bauwerk Parkett have been awarded the Cradle to Cradle Certified® gold certificate. Their lacquer consists of most advanced high-tech chemistry, a masterpiece of complex chemical engineering integrating material health and quality performance. It took time to fully implement gold level matching ingredients and to conduct repetitive quality testing for production and final product quality assurance. For public and contract applications in buildings, rigorous property criteria are required. Formulations needed to be updated, and gold level was not feasible, but Cradle to Cradle Certified® certification at silver level was awarded. It makes sense to treat the top layer of the wooden parquet product with natural oils made from renewable resources. The third group of products with Cradle to Cradle

Certified® from Bauwerk Parkett are naturally-based oil lacquers at bronze level, due to inherently emissions carried at very low concentrations.

Essentials: PU Flooring (Polyurethane) - Biobased Chemistry

Polyvinyl Chloride (PVC), also known as vinyl, is an economical and versatile thermoplastic polymer. It is widely used in the building and construction sector for crafting door and window profiles. Vinyl is also found in drinking and wastewater pipes, wire and cable insulation, medical devices, and flooring. It is the world's third-largest thermoplastic by volume, forecast to reach 60 million metric tons in 2025. China is the world's largest PVC producer, with about 80 percent of its total 25-26 million mt/year using coal-based carbide as feedstock, with the rest using ethylene-based feedstock.[41]

PVC is favored for its light weight, durability, low cost, and ease of processability. It has become a popular substitute for traditional materials such as wood, metal, concrete, rubber, and ceramics due to its diverse properties. The polyvinyl chloride market has seen substantial growth, driven by the increasing demand within the building and construction sector, particularly in the production of door and window profiles.

Greenpeace claims that "PVC (vinyl) contaminates humans and the environment throughout their lifecycle, during its production, use, and disposal. While all plastics pose threats to human health and the environment, few consumers realize that PVC is the single most environmentally damaging of all plastics."[42] PVC contains dangerous chemical additives, including phthalates, lead, cadmium, and/or organotin.

---

[41] "China's PVC hits record-high on supply concerns due to power-consumption curbs", S&P Global, Accessed June 4, 2024, https://www.spglobal.com/commodityinsights/en/market-insights/latest-news/chemicals/092221-chinas-pvc-hits-record-high-on-supply-concerns-due-to-power-consumption-curbs

[42] "PVC Is Incredibly Harmful to the Environment – Here's Why", Green Matters, Accessed June 4, 2024, https://www.greenmatters.com/p/why-is-pvc-bad-for-the-environment

Within the Cradle to Cradle Certified® certification, PVC is banned—i.e., products that contain PVC cannot be certified.

Alternatives include polyurethane flooring, which is a step into the future. Polyurethane flooring from Windmoeller is a synthetic flooring manufactured primarily using sustainable raw materials and natural fillers with no chlorine, plasticizers, or solvents.

Windmoeller GmbH, a flooring brand, embodies innovative energy, a focus on service, and a friendly working atmosphere. Their core competencies are floor coverings (wineo) and wood-based panel products. With over seventy-five years in the industry, Windmoeller Holding and all its affiliated companies stand for progress through innovative and high-quality products, as well as service for their customers, in more than seventy countries all over the world.[43]

Polyurethane flooring is a sturdy solution for high-traffic areas, thanks to the special polyurethane wear layer. The secret is a material called ecuran. This high-performance composite product is manufactured from plant-based oils such as canola oil or castor oil and natural mineral components like chalk. Together with its extremely high levels of resilience and cleaning ease, ecuran is both sustainable and economical.

However, in 2020, the EU published a new REACH regulation on the production of polyurethanes that require aromatic or aliphatic diisocyanates. These can be handled safely when used in accordance with established safety measures. In the PU production process, the diisocyanates react with polyols and are used up, so that they do not exist in the finished polyurethane products. The restriction does not ban the use of polyurethane adhesives and sealants but establishes requirements for their safe use. PU adhesives and sealants will therefore continue to be

---

[43] "Innovative Floor Coverings", Windmöeller GmbH, Accessed June 4, 2024, https://www.windmoeller.de/en/solutions/floor-coverings/

widely used because of their versatility and because no other technologies exist with all the properties PU has.[44]

Essentials: Rubber Flooring - Natural Rubber – Inherited Industrial Knowledge

The farming and production of natural rubber has had devastating social and environmental impacts, with large forest clearances to cultivate plantations over fourteen million hectares worldwide. Like palm oil, rubber plantations are usually monocultures, degrading the habitat, biodiversity, soil, and polluting the local environment with chemicals. Around six million smallholder farmers produce around 85 percent of the world's natural rubber. Thus, ensuring fairness, transparency and traceability for rubber cultivation and rubber processing is difficult.

Latex is used for many products, such as gloves, wound drains, urinary catheters, hot water bottles, Ambu bags, blood pressure cuffs, dental wedges and rubber dams, dressings, rectal tubes, and so on. Manufacturing efficiency, user comfort, and low price make latex a popular choice. However, proteins in latex can trigger allergies.[45]

A Cradle to Cradle® project was set up by Artigo, a rubber flooring manufacturer in Italy, with the inherited industrial knowledge of Pirelli. Artigo offers innovative products that stem from its research within the Pirelli Group in the 1920s.

The aim of the project was to acquire Cradle to Cradle Certified® certification and achieve silver level. The company also wanted to tailor its perspective on material health, product circularity, clean air and climate protection, water and soil stewardship, and social fairness. It's not an

---

[44] "A safe future for polyurethane products", European Coatings, Accessed June 4, 2024, https://www.european-coatings.com/news/legislation/a-safe-future-for-polyurethane-products/

[45] "Latex allergy", aha! Swiss Allergy Centre, Accessed June 4, 2024, https://www.aha.ch/swiss-allergy-centre/allergies-intolerances/latex-allergy?lang=en

easy task to transform this industry, to close the loop to reintegrate raw materials back into the supply chain.

Artigo has partnered with Mondo, the world leader in rubber applications for business and the sports industry. These two industrial cultures together produced a vast and diverse collection, with an exceptional number of different applications.[46]

---

[46] "Artigo", ERFMI, Accessed June 4, 2024, https://erfmi.com/portfolio-items/artigo-s-p-a-mondo-group/

CHAPTER 11

Transforming Systems - From Linear to Circular Systems - Case Studies

Concept Framing in Countries: The European Green Deal (EU Commission)

Period: 2007 - 2017

Focus: Legislation and Political Framework for Sustainability

The European Green Deal, initiated by the EU Commission, represents a comprehensive strategy to transform the European economy towards a more sustainable and circular model. This ambitious plan aims to improve citizens' well-being and health through multiple initiatives like ensuring clean air and water, promoting energy-efficient buildings, and encouraging sustainable food systems. The Green Deal is envisioned as Europe's "man on the moon moment," symbolizing a significant leap in sustainable development. It encompasses extensive legislative changes, driving companies to adapt swiftly to remain compliant.

Consumer Behavior: Migros

Period: 2006 - 2021

Focus: Retail Transparency and Circular Product Initiative

Migros, Switzerland's largest retailer and private employer, has made substantial strides in sustainable retail. Their commitment to transparency is evident in their M-Check system, which rates products on sustainability criteria like animal welfare, fish stewardship, climate compatibility, packaging, material health, and product circularity. This initiative has created a "pull effect," urging suppliers globally to transform their products into safe and circular offerings. Migros's approach highlights the influential role of retailers in shaping consumer behavior and promoting sustainable products.

DNA Buildings + Facility Management, Circular Accounting by epeaswitzerland™

Focus: Construction and Facility Management

Circular Accounting by epeaswitzerland™ for Buildings & Facility Management
DNA Buildings + Facility Management

Fiscal year

		Aluminum	Aluminum	Aluminum	Investment Property Steel	Steel	Steel			Total
		C2C > 20 years	C2C < 15 years	Recycling	C2C > 20 years	C2C < 15 years	Recycling			
1,0	Evaluation rate in %	80%	80%	30%	80%	80%	30%			
	Quantities in the building	200 t	80 t	35 t	75 t	150 t	88 t			
	Exchange rate US$ - CHF	0,90	0,90	0,90	0,90	0,90	0,90	0,90	0,90	
	Value of raw materials/ton of acquisition in CHF	1.500,00	1.500,00	1.500,00	300,00	300,00	300,00			
	Value of raw materials / ton currently in CHF	1.700,00	1.700,00	1.000,00	350,00	350,00	200,00			
2,0	Book value at the beginning of the financial year	300.000	120.000	52.500	22.500	45.000	26.400			566.400
3,0	Additional Entries	0	0	0	0	0	0			0
4,0	Additional Exits	-	-	-	-	-	-			0
5,0	Book value before depreciation	300.000	120.000	52.500	22.500	45.000	26.400	0	0	566.400
6,0	Rawmaterial commodity value adjustment up to date	272.000	108.800	10.500	21.000	42.000	5.280			459.580
7,0	Past added value taxed									
8,0	Value added purchase according accounting income statement									0
9,0	Value before depreciation	300.000	120.000	52.500	22.500	45.000	26.400	0	0	566.400
10,0	Booked depreciation value									
10,1	via inventory account	0	0	0	0	0	0	0	0	0
10,2	Booked value added purchase according accounting income statement	0	0	0	0	0	0	0	0	0
10,3	Total depreciation	0	0	0	0	0	0	0	0	0
11,0	Regular Depreciation							0	0	0
12,0	Additional depreciation	0	0	0	0	0	0	0	0	
13,0	Increase/decrease in taxed VAT									
14,0	Surcharge compensation	in %	0%	0%	0%					
15,0	Updated taxed VAT	0	0	0	0	0	0	0	0	
16,0	Book value after depreciation	300.000	120.000	52.500	22.500	45.000	26.400	0	0	566.400
17,0	Properties market value									
18,0	Real estate final value									

© epeaswitzerland gmbh

Figure 13: Circular Accounting by epeaswitzerland™ for construction business.

Circular accounting integrates circular principles in the construction and facility management sectors. The approach aligns with the Cradle to Cradle® DNA for Buildings and Facility Management™, advocating for the use of non-toxic, recyclable materials in construction. Our initiatives emphasize the importance of Circular Accounting by epeaswitzerland™ in managing building resources, highlighting the shift from traditional linear models to sustainable circular systems in construction and facility management.

To prove economic viability for buildings and facility management, conventional linear accounting has to be replaced by Circular Accounting.

Integrated Management Systems for Circular Product Design

Designing circular products necessitates an integrated management system approach. This means considering the end-of-life of products at the design stage itself. Products should be designed for durability, reparability, and recyclability without leaving any toxic residuals in recyclates or from processes. An integrated management system also involves collaboration across various departments and stakeholders to ensure that every aspect of the product's life cycle is considered for circularity.

Reintegrating Materials into the Supply Chain

The core of a circular system is the ability to reintegrate materials back into the supply chain. This involves creating efficient collection and recycling systems that can turn waste back into valuable, toxic-free resources. Companies need to establish systems for the collection of used products and invest in technologies that can efficiently recycle these materials safely to virgin material quality level.

Overcoming Barriers to Circular Transformation

One of the main challenges in transitioning to circular systems is overcoming existing infrastructural and technological barriers. This requires significant investment in new technologies and processes. Additionally, there's a need for supportive policies and regulations that encourage circular practices and provide incentives for companies that adopt such practices.

The Role of Collaboration and Innovation

Collaboration across industries and sectors is crucial for the successful implementation of circular systems. Sharing knowledge and best practices can accelerate the transition. Furthermore, innovation is key, both in terms of technology that enables circularity and in business models that support the circular economy.

Transforming systems from linear to circular is a complex but essential journey for sustainability. It requires an approach which encompasses changes in design philosophy, production processes, consumer behavior, and policy frameworks. As we move forward, it's clear that the path to sustainability lies in our ability to close the loop, creating systems where waste is not an end product but a valuable resource for new beginnings.

As a changemaker, my focus is on advocating for and implementing these principles of circularity. The shift from linear to circular systems is not merely an environmental imperative but also an economic and social necessity.

Industrial Composting, Navigating Challenges and Innovations

Biomass is shrinking globally, and agriculture is not able to stop it. Therefore, the Cradle to Cradle® principles envisioned over thirty years ago are asking for industrially produced products to be safe for biological cycles, so they can return to soil, becoming a nutrient for new life. The challenges are significant. Products need to be economically feasible and approved by authorities. It remains a challenge to bring different cultures, industry, supply chain, waste management, agriculture, authorities, science, certification bodies on the same page.

The Role of Authorities in Industrial Composting

In industrial composting, the primary role of authorities is crucial to prevent worst-case scenarios, such as environmental contamination and public health risks. This requires stringent regulations and monitoring systems to ensure that composting facilities operate within safe and sustainable parameters.

Innovations in composting often require acceptance and approval from authorities, particularly when they involve new technologies or methodologies. Gaining this acceptance can be challenging, as it involves demonstrating the safety, efficiency, and environmental benefits of new composting approaches. Collaborative efforts between innovators and regulatory bodies are therefore essential.

While authorities focus on preventing negative outcomes, innovation in industrial composting aims to achieve composting efficiency and environmental benefits. The challenge lies in aligning these innovations with regulatory frameworks, ensuring that new composting technologies are both effective and compliant with environmental standards.

Overcoming Technical and Economic Challenges in Industrial Composting

For industrial composting to be viable, it needs to be economically sustainable. Achieving a break-even point is essential for businesses to invest in composting technologies. This involves optimizing processes to reduce costs and finding market opportunities for compost products. Economic sustainability also encourages more businesses to adopt composting practices, contributing to broader environmental goals.

Advances in industrial composting can be stymied by technological limitations, lack of infrastructure, and regulatory hurdles. Overcoming these challenges requires a multi-faceted approach that includes technological innovation, policy advocacy, and public-private partnerships. Educating stakeholders about the benefits of composting and developing incentives for composting initiatives can also play a significant role.

The Future of Industrial Composting

Looking ahead, the future of industrial composting lies in integrating this sector seamlessly into waste management systems. This includes expanding composting facilities, enhancing community participation, and developing markets for compost products. Advances in technology, such as automated sorting and processing systems, can further enhance the efficiency and scalability of composting operations.

Industrial composting is a critical component of the circular economy, turning organic waste into valuable resources. The goal is to create systems where waste is a resource, contributing to environmental sustainability and circular economy principles.

The Circular Economy and the New Green Deal

The New Green Deal: Europe's Ambitious Plan

The European Green Deal, championed by Ursula von der Leyen, represents a bold initiative to make Europe the first climate-neutral continent by 2050. This political framework is significant in driving legislation that aligns with environmental objectives, balancing ambitious goals with a careful assessment of impacts.

Integration into Legislation

Legislation to support the Green Deal is a complex but necessary step to make environmental goals not just aspirational but actionable. Such laws and regulations will ensure that all sectors contribute to the sustainability targets. It's a comprehensive approach, impacting different interest groups and lobbyists from industries, communities, buildings, agriculture, resources, waste management, water and energy sectors, logistics, transportation, and beyond.

Public Hearing on "Traceability of Problematic Substances in Recycling"

A public hearing on "Traceability of Problematic Substances in Recycling" in 2018 underscored the need for systems that can track and manage substances that pose a threat to sustainability. Such monitoring is crucial for preventing the reintroduction of hazardous materials into the supply chain.

The Circular Economy and the New Green Deal represent critical steps towards a sustainable future. They embody a shift in thinking and policy-making that prioritizes the health of our planet and its inhabitants

Transforming the Textile Industry in Indonesia: A Journey to Circularity and Cradle to Cradle®

First Project in Developing Countries

For my first major project in a developing country, I went to Indonesia to transform the textile industry. This project was not just about business efficiency, but also instilling sustainability and circularity in a sector deeply rooted in the country's culture and economy.

Circularity and Cradle to Cradle® in Textile Industry

The overarching vision of the Indonesia project was to revolutionize the textile industry throughout in this developing country, transitioning it from traditional linear models to a circular, sustainable framework. The goal was both to change production practices and embed the Cradle to

Cradle® philosophy deep within the industry's DNA. This meant designing products and systems in which materials were continually reused, thus reducing waste and conserving resources.

Our approach began with a comprehensive feasibility study to understand the landscape of the textile industry in Indonesia. This involved analyzing existing production processes, supply chains, waste management systems, and environmental impacts. We also engaged with local textile engineering universities and organizations to establish a foundational network for collaborative change.

Challenges and Opportunities

The project encountered significant challenges, particularly regarding infrastructure, technology, and funding. However, funding was the primary hurdle, posing a major impediment to the proposed changes. The infrastructural changes were monumental, requiring significant investment and technological upgrades.

Although the transformation faced several challenges, it also presented significant opportunities for innovation and sustainable growth. We collaborated with local stakeholders and leveraged global best practices to create a model that could be replicated in other developing countries.

The scale of the project required partnerships across various sectors. This included working with local communities, industry leaders, government bodies, and international organizations. The aim was to create a united front, pooling resources, knowledge, and expertise.

Cultural and Economic Considerations

Respecting and integrating Indonesia's rich textile heritage and cultural practices was an important part of the project. We aimed to balance sustainability with the preservation of traditional methods. The transition had to support local artisans and workers. Economically, the project aimed to demonstrate that sustainability could coexist with profitability.

Impact and Future Outlook

The transformation of Indonesia's textile industry was more than a local initiative. It provided a blueprint for global sustainability. The project demonstrated how collaboration, innovation, and commitment can reshape industries towards a more sustainable and equitable future.

The model intertwined the preservation of traditional skills with modern, eco-friendly methods. It showed that progress need not come at the cost of cultural identity.

This transformative project was not just environmental conservation. It represents a broader vision for the future of global industries, especially in developing countries.

Transitioning to Circular Cleaner Production

Defining Cleaner Production

"Cleaner Production" is a preventative approach in environmental management. The focus is on minimizing waste and emissions at the source rather than dealing with them after production. This concept places environmental concerns at the core of industrial processes.

Traditionally, Cleaner Production follows a linear approach, where the primary goals are to reduce, minimize, and manage waste and emissions efficiently. This model operates on a 'beginning-to-end' principle, starting with raw material extraction and ending with waste management. This approach has been effective in reducing environmental impacts to a certain extent. However, waste and emissions are treated as unavoidable by-products, rather than resources that can be reintegrated into the production cycle.

The efficiency goals of Cleaner Production in a linear model focus on reducing resource use and minimizing waste generation. This only leads to incremental improvements rather than transformative changes. The emphasis is on doing less harm rather than creating positive environmental impacts. The linear model does not fully address the root causes of environmental degradation.

Redefining the Production Approach

To overcome the limitations of the linear model, there is a growing shift towards integrating Cleaner Production with circular economy principles. This involves rethinking and redesigning production processes to ensure that materials and products are kept in use for as long as possible. In this closed-loop system, waste is minimized, and resources are continuously recycled.

Achieving circularity in Cleaner Production requires collaboration across the entire value chain, from suppliers to consumers.

Cleaner Production, when aligned with circular economy principles, offers a transformative approach to industrial processes.

Safe and Circular Products and Closing the Loop

Not-for-Profit Independent Certification

Not-for-profit, independent certifications promote certain standards and practices, often related to environmental sustainability, ethical business practices, and social responsibility. My advocacy for such certifications stems from a deep commitment to sustainability and my belief that transformative change in industries is both necessary and achievable. These certifications are a powerful tool in this journey towards a more sustainable future.

The broader impact of these certifications extends beyond environmental benefits. They also contribute to raising consumer awareness about sustainable practices, influencing purchasing decisions, and driving the overall market towards more sustainable options.

Certification not only sets standards for safe and sustainable products, but also help change mindsets, promote responsible consumption, and facilitate the transition towards closing the loop in various industries.

The Importance of Independence

The independence of certifications guarantees an unbiased and objective assessment of compliance with set standards. The certification process is free from any conflicts of interest, particularly from commercial motivations.

Changing the Mindset and Bringing Together Competitors

One of the key roles of Not-for-Profit Independent Certifications is to foster a shift in mindset among producers, consumers, and other stakeholders. By setting and upholding high standards, they encourage movement from traditional, often unsustainable practices to more responsible and environmentally friendly approaches.

One of the most significant challenges and achievements in building a consortium network is bringing together competitors. This requires a shift in mindset from viewing competitors solely as business rivals to seeing them as partners in achieving a greater good. The focus is on shared value creation rather than individual gains.

Strategies for Effective Collaboration

Effective collaboration in a consortium network involves clear communication, shared objectives, and a governance structure that allows for equal participation and decision-making. It's about leveraging each company's strengths and resources for collective innovation.

Knowledge Sharing and Innovation

One of the key benefits of such networks is the pooling of knowledge and resources. This facilitates innovation, especially in areas like material science, manufacturing processes, and product design, which are crucial for developing safe and circular products.

Overcoming Challenges

Building and sustaining a successful consortium network is not without challenges. These include aligning different corporate cultures, managing competitive sensitivities, and ensuring equitable contributions and benefits among members. Overcoming these barriers requires a strong foundation of trust and a shared vision of sustainability.

Impact on Circular Economy

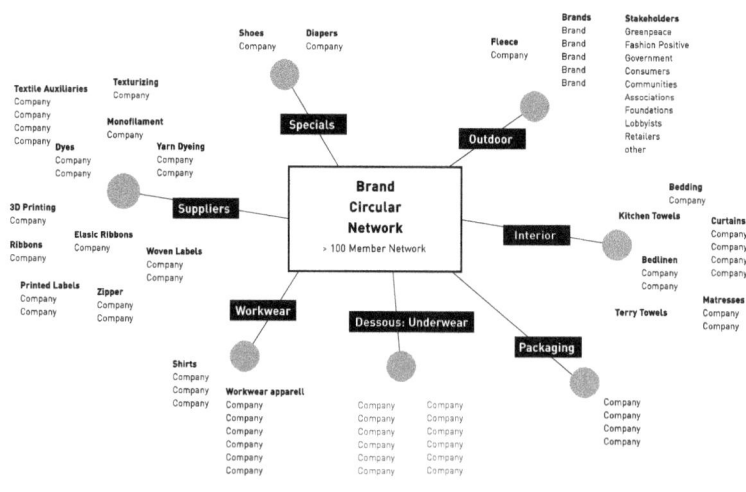

Figure 14: Transforming economic with circular networks.

Consortium networks play a critical role in advancing the circular economy. By collaborating, companies can accelerate the development of sustainable solutions, create economies of scale, and influence broader industry and regulatory changes.

Building a consortium network, as exemplified by Wolford and others, is a testament to what can be achieved when companies, even competitors, unite under a common sustainable vision. As we move forward, the role of such networks will be increasingly pivotal in transitioning industries towards more sustainable and circular practices.

Real economic values can be generated within the network of suppliers, brands, consumers, take back logistics companies and the reintegration of raw materials back into the supply chain.

From Vertically Integrated to Circular Integrated Company

In 2019, Sanko textiles took on the challenge of transforming from a vertically integrated to a circular integrated company. The company began establishing close partnerships with agriculture to grow organic and regenerative cotton and invested in recycling technologies to reintegrate fibers back into the supply chain.

Since its founding in 1904, Sanko Textile has grown as the cornerstone of Sanko Holdings, one of the strongest corporations in Turkiye. SANKO Holdings employs over fourteen thousand persons and includes internationally renowned companies in textiles, energy production, construction, and construction machinery. "True success is being honest in Sanko's every action, including those towards nature, people and all other creatures on our planet," the company says.

The Need for a Cooperative in Fashion

The fashion industry often faces challenges in accessing sustainable materials and resources. This is especially so for startups. Establishing a cooperative is essential to provide startups with the resources they need to innovate and create environmentally friendly products.

One of the primary goals of a cooperative would be to ensure that startups have access to materials that are safe for consumers and the environment, and circular by design.

Support and Collaboration

The cooperative would also provide knowledge sharing, technical expertise, and market access. By bringing together various stakeholders, including material suppliers, designers, and environmental experts, a cooperative fosters a collaborative environment for sustainable innovation.

Advantages for Startups

Startups, often limited by resources and reach, can significantly benefit from this approach. It provides them with a platform to experiment with sustainable materials, gain insights from industry experts, and collaborate with other like-minded entrepreneurs.

Long-Term Goals

By empowering startups, the cooperative incubates the next generation of fashion innovators who are committed to environmental stewardship.

This approach reflects my belief in the power of collaboration and innovation in driving sustainable change. By supporting startups in this manner, we can make the fashion industry more aligned with the principles of circularity and sustainability.

Concept Framing for Countries: EU Commission's Green Deal

Introduction to the EU Green Deal

The European Green Deal, initiated by the EU Commission, is a comprehensive plan to make the EU's economy sustainable. This ambitious policy aims to transform the EU into a fair and prosperous region, with a resource-effective economy, with no net emissions of greenhouse gases by 2050, and where economic growth is decoupled from resource use.

Legislation and Political Momentum

A key aspect of the Green Deal is binding legislation. This rapid political movement signifies a substantial shift in how environmental policies are integrated into the broader economic and social framework of EU member countries. It reflects a decisive step away from voluntary measures towards legally enforceable standards and practices.

Impact on Companies

The legislative nature of the Green Deal carries significant implications for businesses operating within the EU. Companies will need to adapt swiftly to these changes to ensure compliance with new legal standards. This includes revamping processes, investing in new technologies, and rethinking business models to align with the EU's sustainability goals.

Compliance and Legal Implications

For companies, the fear of non-compliance isn't just about penalties or fines—it's about maintaining their license to operate in an increasingly environmentally conscious market. Non-compliance may lead to legal challenges, as well as reputational risks and loss of consumer trust. Hence, businesses have a pressing need to proactively align with the Green Deal's objectives.

Opportunities in the Green Transition

While the Green Deal presents challenges, it also opens numerous opportunities for innovation and leadership in sustainable practices. Companies that embrace these changes can gain a competitive advantage, access new markets, and benefit from a positive brand association.

The Green Deal's transformation into legislation signals a collective move towards a sustainable future and sets a precedent for environmental policy globally. For companies, this shift requires foresight, agility, and a commitment to sustainability, not just as a compliance measure, but as a core business strategy.

The EU Commission's Green Deal represents a significant opportunity for businesses to lead in the transition to a regenerative future. The move from voluntary to mandatory compliance underscores the urgency of environmental issues and presents a clear directive for companies to innovate and adapt.

The Green Deal is more than just a policy; it signifies a new era of environmental and societal responsibility and offers a roadmap for a sustainable future.

Understanding System Integration in a Company

System integration in a company involves aligning and coordinating various components of the business, from production processes to management decisions, to work seamlessly towards common goals. This includes integrating sustainability and environmental responsibility into every aspect of operations.

The Role of Materials and Decisions

In the context of sustainable practices, every material used, and every management decision made plays a crucial role. For instance, the choice of materials, like adhesives in production processes, must align with sustainability, as well as material health and circularity principles. The decision to use certain types of glue not only impacts the product's environmental footprint but also its compliance with sustainable or material health or circularity certifications.

Marketing Challenges: The Case of Bauwerk Parkett, "Invisible but Measurable" Challenge

Bauwerk Parkett's marketing challenge of "Invisible but Measurable" highlights a common issue in promoting sustainable products. While the environmental benefits of sustainable practices like using eco-friendly glue might be significant, they are often not immediately apparent or tangible to consumers.

Communicating Sustainability

The challenge lies in effectively communicating the value and impact of these sustainable practices to customers. This requires innovative marketing strategies that can convey the environmental benefits and the added value of products that are Cradle to Cradle Certified®.

Leveraging Certification in Marketing

Using Cradle to Cradle Certified® Certification as a marketing tool involves highlighting the rigorous standards the products have met and the tangible environmental benefits they offer. It's about educating consumers on the importance of sustainability and how choosing certified products contributes to a larger environmental cause.

The Challenge of Circularity in the Electronic Industry

Current State of the Electronic Goods Industry

The electronic goods industry, despite its rapid growth and innovation, has been slow to move towards circularity. A decade after the initial push for circular economy practices, the industry still largely operates on a linear model, producing, using, and disposing of electronic goods.

Barriers to Circularity

One of the main challenges in transforming the electronic goods industry is the complexity and rapid technological advancement inherent to this sector. Products often become obsolete quickly, leading to high turnover and waste. Additionally, recycling electronic components is complex due to the variety and kinds of materials used and the difficulty in disassembling products.

Creating Circular Business Models

Design for Longevity and Repairability

To transition towards a circular model, businesses in the electronic industry need to rethink product design. This includes designing for longevity, where products are built to last longer, and for repairability, where products can be easily disassembled and repaired, extending their life cycle.

Modular Design and Upgradability

Another strategy is modular design, where products are designed in a way that allows for easy upgrades of specific components instead of replacing the entire device. This not only reduces waste, but also takes into account the rapid pace of technological advancements.

Encouraging Product-as-a-Service Models

Shifting from product ownership to service-based models can also drive circularity. In such models, companies retain ownership of the products and provide them as services to consumers. This encourages manufacturers to produce more durable and repairable products, as they are responsible for their maintenance and eventual recycling.

Implementing Take Back and Recycling Programs

Effective take back and recycling programs are essential for closing the loop in the electronics industry. By ensuring that old electronics are collected and responsibly recycled, valuable materials can be recovered and reused, reducing the need for new raw materials.

Leveraging Technology for Traceability

Advancements in technology can be leveraged to improve traceability in the supply chain. By tracking the origin and journey of materials, companies can ensure responsible sourcing and optimize the recycling process.

Consumer Awareness and Participation

Consumer awareness and participation are critical in driving the shift towards circularity. Educating consumers on the environmental impact of electronic waste and encouraging them to participate in recycling and take back programs is vital.

Transforming the electronics industry requires a multifaceted approach that includes innovative product design, business model shifts, and effective recycling programs. A circular electronic goods industry has vast benefits, including reduced environmental impact, conservation of resources, and the creation of sustainable value chains. The adoption of circular economy principles in the electronic industry requires business models that not only address current environmental challenges, but also set the foundation for a sustainable and resilient electronic goods industry.

Consumer Behavior in the Context of Retail Transparency and Circularity

The Shift Towards Transparency in Retail

In recent years, there's been a significant shift in consumer expectations towards transparency in the retail sector. Consumers increasingly demand to know more about the products they purchase, from their origin and manufacturing processes to their environmental impact.

The Concept of Migros, Switzerland, M-Check in Retail

M-Check is owned by Swiss retailer Migros and pre-evaluated star rating on material health and circularity of their nonfood products. This involves evaluating how products are sourced, produced, and eventually recycled or disposed of. Retailers who implement M-Check are committing to high standards of environmental responsibility and consumer safety.

Consumer Response to Transparency

Consumers are becoming more environmentally conscious and are willing to support retailers who demonstrate a commitment to sustainability. Full transparency, as represented by the adoption of M-Check, empowers consumers to make informed choices that align with their values. This transparency fosters trust and loyalty among consumers towards retailers.

This shift towards more sustainable and transparent retailing has a significant impact on consumer behavior. Consumers are more likely to support brands that demonstrate a commitment to sustainability, leading to a change in purchasing patterns. They tend to prefer products that are not only safe for use, but also contribute positively to environmental sustainability.

The Pull Effect on Global Suppliers

Retailers who commit to transparency and circularity create a pull effect in the supply chain. Their commitment encourages or necessitates suppliers worldwide to transform their products to meet these standards. Suppliers are incentivized to adopt safe and circular practices to maintain their business relationships with these retailers.

Challenges and Opportunities

One big challenge for retailers is the lack of recycling infrastructure for many products. The move towards transparency and circularity presents challenges for retailers and suppliers, such as the need for systemic changes and potential increased costs. However, it also offers significant opportunities. Retailers who successfully integrate these practices can differentiate themselves in the market, attract environmentally conscious consumers, and potentially tap into new customer segments.

This shift represents a significant step towards a more sustainable and responsible retail industry (see Chapter 14).

CHAPTER 12

Transforming Management - Management Tools for Circularity - Case Studies

As an innovator and advocate for the circular economy, I have long recognized the need to move beyond traditional linear management and accounting practices. Conventional methods, deeply ingrained in our economic systems, do not align with the principles of circularity, which focus on resource sustainability and long-term viability.

Circular Accounting by epeaswitzerland™

The core issue with traditional accounting is its linear nature, which fails to capture the true economic essence of circular processes. In circular accounting, the value of resources is continuously accounted for. This includes considering the costs and value of raw materials throughout their lifecycle, not just as a one-off transaction. This fundamentally changes the economic factors of raw materials.

For instance, in circular accounting, if a product is designed based on the Cradle to Cradle® concept, its raw materials maintain their quality and non-toxicity, allowing them to be safely reintegrated back into the supply chain. This fundamentally changes the economic valuation of these resources compared to linear accounting where, once a product reaches its 'grave,' its material value is typically disregarded.

Transparency and Cost Structure: a critical aspect is creating transparency in financial aspects. This transparency enables us to illustrate the cost structure over multiple life cycles, demonstrating how products become cheaper from cycle to cycle as they are reused and cycled over and over again.

Challenges with GDP in a Circular Economy

The concept of Gross Domestic Product (GDP) is another area where traditional economic metrics fall short in a circular economy. GDP is designed to measure the growth of a linear economy, focusing on production and consumption. However, in a circular economy, the emphasis shifts to resource efficiency, reuse, and long-term sustainability. Therefore, the GDP might not accurately reflect the health or success of an economy transitioning towards circularity. We need more tools that incorporate the principles of circular economy.

Transparency in Products

Transparency in products is becoming increasingly important. The shift from legal requirements of 1000 parts per million (ppm) to 100 ppm signifies a move towards full disclosure of product contents. This level of transparency is essential in a circular economy. It allows companies and consumers to understand and manage the impact of products on health and the environment. This shift is not only a legal compliance issue, but also builds a toxic-free environment.

Leadership and Innovation in Sustainability

My experience as the CEO of a textile mill in Switzerland, which became the subject of case studies at IMD Business School, reflects the importance of leadership and innovation in sustainability. These studies, titled "Surviving the Impossible" and "Leveraging Sustainability," showcase how a small company can overcome significant challenges through innovative sustainability strategies. The Climatex product, a Cradle to Cradle Certified® certified textile, is a testament to this approach.

Building the Team: Aligning Individual Values with Circularity Goals

Aligning team members' personal values with circularity goals is crucial for success. This alignment can be achieved by discussing values and creating a shared understanding and respect among team members. Such alignment brings everyone to the same level of understanding and commitment, which is essential for driving circular initiatives. Building an effective team for circularity initiatives involves more than just gathering a group of individuals with the right technical skills. It's about aligning each team member's personal values and motivations with the overarching goals of circularity. This alignment is not about missionary zeal or brainwashing approaches; it's about making sense based on individual values and motivations.

Supply Chain Domino-Knowledge Transformation™

The supply chain domino-knowledge transformation™ is a concept we developed to accelerate the transition to circularity. By bringing together the entire supply chain in one meeting and communicating requirements efficiently, this approach significantly reduces project time. This was exemplified in our work with Napapijri, where we brought the network together and achieved quick prototyping, setting a new standard for project execution.

Management Tools for Circularity

Finally, the development of management tools for circularity is essential. Managers and business students need to think and operate differently. The case studies and concepts presented in my work enables management to implement circular economy principles effectively in real-life business scenarios.

Transforming management for circularity is about redefining traditional concepts like accounting, fostering transparency, encouraging positive impacts, scope of design and driving innovation. It's about creating a shared vision, aligned with individual values, and building a network of trust by epeaswitzerland™ that redefines how we value resources and manage our operations.

The Approach:

1. **Reference Model Cradle to Cradle®**: We begin by establishing a common framework, the Reference Model Cradle to Cradle®, which consists of eighteen points, akin to a golf course. This model serves as a guide to discuss and align key aspects of circularity.
2. **Values Discussion**: Each team member is encouraged to articulate their values and how they see the future. This can range from broad goals like making the world safe for future generations to everyday actions like commuting to work by bicycle.
3. **Creating Respect and Understanding**: By sharing these personal values, team members develop a deeper respect and understanding for one another. This shared understanding helps create a level playing field where every opinion and value is respected.
4. **Applying Values to Work**: In one workshop I conducted for a company specializing in outdoor furniture, we started by placing various materials on the table. These materials represented the company's current usage. After a session of values sharing, we had each participant select a material they believed was the future for the company. Remarkably, each material was valued equally, demonstrating the team's unified vision and respect for different perspectives.

This approach underscores that successful circularity initiatives are not only in innovative business models or technical expertise, but also in personal values and motivations of the people involved.

Marketing Based on Science: Avoiding Greenwashing

The transition to a circular economy also demands a revolution in how we approach marketing. It is crucial to base marketing strategies on solid scientific facts to avoid greenwashing—i.e., making unsubstantiated or misleading claims about the environmental benefits of a product.

1. **Scientific Foundation**: The key to avoiding greenwashing lies in incorporating scientific insights into marketing. We ask scientists to provide statements on material health and circularity. These statements form the backbone of our marketing narratives.
2. **Storytelling Based on Facts**: The marketing teams take these scientifically grounded statements and weave them into compelling stories. These stories are not flowery or extravagant claims but rooted in verifiable facts.
3. **Case Example**: In a project for wooden flooring, the challenge was to authentically communicate the environmental credentials of the product. The marketing team came up with the concept of "Invisible but Measurable," emphasizing that, while the benefits of the product might not be immediately visible, they are quantifiable and significant. This approach ensures that consumers receive factual information, contributing to building trust and credibility in the brand and its commitment to circularity.

In marketing, avoiding greenwashing through science-based storytelling is crucial for credibility and trust in the circular economy. These strategies are essential for any company aiming to transition to a sustainable, circular business model.

CHAPTER 13

Embracing Startups and Their Potential - Case Studies

Start-Ups - Proof of Concept

Essentials:

Not only have we been educated to think in a linear manner, but all our systems are based on linear thinking. System barriers are all over the place. Finding companies that are willing to adopt a new mindset is not easy. When it comes to products that dominate the market, it is almost impossible to inspire a hero or pioneer or a company to stand up and make the impossible possible.

On the other hand, there are ambitious entrepreneurs who want to change something, who usually have hardly any money, but have a good idea and the will to break the wall. What do they do? They found a start-up company.

Start-ups have the advantage of starting from scratch. They have no historical baggage and no boundaries, and the freedom to take risks. They are driven by innovation and are committed to their beliefs.

At some point in life, you need to give something back to society. This ambition has been my personal concrete desire for many years, but to find the right vehicle is not that easy. As my conviction matured over the years, I decided that support for start-ups would provide that opportunity. Doing this would also accelerate the transformation of the linear economy into a circular economy, perhaps even leading to the creation of novel products that would overcome the resistance of market leaders. The driver is to "prove the concept."

As the 100 percent owner of epeaswitzerland gmbh, I can take full advantage of making decisions fast and independently, but I have to bear in mind all the consequences of doing so. I decided to offer to specific, committed start-up companies' services that are rendered by a professional for free or at a lower cost.

Beverages Packaging

BAYONIX®

Polyethylene terephthalate (PET) is the most widely used material in the beverage bottling industry worldwide. The production of PET bottles is estimated at more than five hundred billion per year, which is approximately one million bottles every minute. However, the use of PET has significant harmful effects on the environment, creating microplastics with no biodegradability. As noted earlier, antimony trioxide, which is classified as possibly carcinogenic, is used as a catalyst in the production process. Approaches to mitigate the negative environmental impact of the PET bottle industry to lower or reduce the issues are designed to conceal symptoms such as reducing the amount of material used and designing for recycling and reuse.

For decades, I have tried to contribute circular and Cradle to Cradle® Design solutions of PET in textiles, as well as drinking bottles. The resistance has been enormous. A lobby organization for antimony was founded in Brussels, and NGOs negated the antimony issue at public events where I presented and talked about the issue. Regulators have classified PET bottles as food-grade approved.

Then I met Stefan Hunger. He was the brother of a good friend of my wife's son, and he wanted to do something meaningful in life. Stefan was on fire with the idea of developing beverage bottles to solve the environmental issues holistically, such as microplastics that were safe for biological cycles, recyclability, and meeting the criteria for circular product properties.

In other projects, we had come across a biodegradable polymer, which was successfully used to produce fibers or extrusion. However, plastic molding technology was a challenge. Stefan contacted about eighty plastic molding companies before finding one producer who had the right machinery and was willing to do some trials. This was no easy task, especially since doing trials on production machines can be costly if they fail or damage the tooling equipment. Eventually, the efforts were successful and the parameters for production were defined. Financing the tooling equipment was challenging for a start-up company, but Stefan succeeded and was able to start production and marketing under the trademark BAYONIX®. However, this market is highly competitive, and who wants to buy a product with novel properties? One major obstacle was that the plastic safe for biological systems couldn't handle temperatures higher than 50 degrees Celsius. This meant that, because of the imprecise temperature controls of dishwashers, BAYONIX® had to label the bottles as not dishwasher safe. We didn't think this would be a problem, believing that environmentally conscious consumers would accept this in the same way they accept that you can't put wooden or silver cutlery in the dishwasher. It turned out that consumers were not willing to take this extra step with the bottle. This was an unfortunate and disappointing experience for us.

Even after a long lifespan, the first beverages drinking bottle with "upcycling premium" has to be disposed of. But you can do so with a clear conscience. The future-proof BAYONIX® Bottle is completely recyclable and safe for biological cycles. BAYONIX® wants a real circular economy, not "Eco-Junk." BAYONIX® products do not become waste! Instead of throwing the bottle in the trash, discarded BAYONIX® bottles, regardless of their condition, can easily be returned back to BAYONIX®. BAYONIX® grants a 15 percent discount on your next purchase in the BAYONIX® online shop, which is delivered by email in the form of a digital voucher. BAYONIX® is working hard on collaborations so that consumers can return the used bottles by other avenues.

BAYONIX® has achieved Cradle to Cradle Certified® certification at gold level. The secret lies in the innovation of the materials, the chemical ingredients, which tested product properties, and recyclability. Production itself is also constantly optimized for sustainability and environmentally friendly criteria, while remaining focused on quality. It's not about repairing and recycling. It's about doing it right the first time. BAYONIX® offers customers high-quality and versatile products at a fair price-performance ratio.

As Stefan says in an official company statement: "I want to show that even the smallest companies can bring economically viable products onto the market that are circular, save virgin resources, and are safe for soil, water, animals and people. The industry often claims that this is not possible."

BAYONIX® firmly believes that sustainability and high-tech are not mutually exclusive but must complement each other. BAYONIX® spends a lot of time and development work every day making their products even better to ensure customer satisfaction. The goal is to develop products that go beyond standards, because standards are the death of innovation. [47]

Cosmetics

Lanz Natur AG

From luxury creams and fragrances to eco-friendly and natural alternatives, the global cosmetics market is seeing an incredible surge in demand. Market shares are: Asia - 32 percent, US - 28 percent, Europe - 22 percent, Africa - 10 percent, and Latin America - 8 percent. Major brands are leading this market.

There is a darker side to the beauty industry that often goes unnoticed - greenwashing. This practice deceives consumers into believing that a product is eco-friendly, natural, and safe.[48]

---

[47] "About Us", Bayonix, Accessed June 3, 2024, https://bayonix.com/pages/ueber-uns
[48] "All About Greenwashing and Cleanwashing in the Beauty Industry", With Simplicity, Accessed June 4, 2024, https://withsimplicitybeauty.com/blogs/withsimplicity-blog/what-does-greenwashing-look-like-in-the-beauty-industry

Cosmetics fuel a high demand for natural oils, leading to extensive and intensive mono-cultivation. This harms natural habitats through deforestation, contaminating soil and water through pesticides, fertilizers and microplastics. Heavy metals like lead, arsenic, mercury, aluminum, zinc, chromium, and iron are found in various personal care products including lipstick and whitening creams. [49]

LANUR is a future-safe cosmetics brand. Owned by Karin Lanz, a Swiss actress and TV host, she launched her own cosmetic company based on her personal principles. "I am a staunch believer that Nature knows best," she says in a company statement. "We cannot possess her, but we can borrow and learn from the best. Borrowing from Nature means that we take something and must ultimately give it back, unharmed."

LANUR was one of our start-up projects, and I and epeaswitzerland supported Karin Lanz from the beginning. The start-up benefited from the ABC-X material health assessment, the integration of the Network of Trust by epeaswitzerland™, and the certification Cradle to Cradle Certified® at gold level.

As a provider of Clean Beauty, LANUR relies on circular materials. Clean Beauty only uses 100 percent recyclable materials for packaging. Applying the closed-loop principle, raw materials are no longer wasted, but only borrowed and then returned to Nature. Waste no longer exists with LANUR. LANUR has submitted to the highest scientific certification standard available in the field of circular economy. It was the first cosmetics company worldwide to receive a gold standard rating, including its packaging. Consumers can return the packaging by mail or leave it at one of the beauty partners. All raw materials are documented, declared, and certified.

This is a perfect example of a 'close the loop' system.

Lanz says, "I am proud to have taken revolutionary new and unusual paths for the development of the world's first Cradle to Cradle

---

[49] "Lead and Other Heavy Metals", Campaign for Safe Cosmetics, Accessed June 4, 2024, https://www.safecosmetics.org/health-effects/cancer/

Certified®-certified cosmetics line, and I guarantee with my name that our products are 100 percent what they promise."

Textiles, Fashion

OceanSafe AG

In 2015, Manuel Schweizer, the Category Manager at Moebel Pfister, Department Store for Interiors and Furniture in Switzerland, contacted me for an interview as he was working on his master's thesis on sustainable textiles. I have supported many students in the past, as it is important to give something back, but a business never evolved out of these interactions. Manuel was on the same track as everyone else, having a linear mindset to be sustainable, linear, cradle to grave, believing in the concept of a "natural romantic" approach. But something triggered him during the interview. His perspective changed radically, and he discovered a new value orientation. The change took a somewhat bizarre form because he not only wanted to rewrite his master's thesis, but also implement the concept in his work. What we did not anticipate was that his commitment for change would become a cornerstone of his life. Manuel managed to convince the supervisory board to allocate a budget for a Cradle to Cradle® project. As a result, many products were innovated, achieving Cradle to Cradle Certified® at gold level. Manuel became an important believer in Cradle to Cradle® thinking.

In 2018, Manuel got an offer from the owner of a German company, Deco Design Fuerus, who was about to retire. He wanted Manuel to be his successor and the new owner. Manuel accepted this position and immediately introduced new Cradle to Cradle® projects, which became Cradle to Cradle Certified® at gold level.

In 2019, Manuel founded a startup company called OceanSafe, located in Switzerland and Germany. OceanSafe's ambition is to address and solve the global textile industry's most pressing environmental issues. OceanSafe provides a fast track to a circular textile industry. OceanSafe's materials will replace conventional textile materials globally. Its materials are suitable for existing value chains. As such, they qualify as drop-in technology and are therefore highly scalable. They match the quality and market prices of existing materials.

OceanSafe operates a licensing business model. The company provides licensed technology to brands and manufacturers along the entire textile value chain. Synthetic circular textile materials have the potential to become a game-changer and to replace conventional materials like polyester or cotton. The concept can be quickly implemented in a global supply chain.

Can this startup company break the giant wall and win the turnaround of the global textile and fashion industry? The future will tell, for a better planet.

Circular Clothing Genossenschaft (Cooperative)

The current fashion industry, and particularly the fast-fashion sector, eats up resources. It wears people out, poisons ecosystems, and destroys the livelihoods of millions. According to the Ellen MacArthur Foundation, as much as 60 percent of all clothes become waste within a year of their production, and 40 percent of all clothes are not even sold or used. This is a tremendous waste of resources. Additionally, many of the materials used in the industry are not safe for people or the environment due to toxic and cancerous substances. These challenges are pervasive in the fashion industry. Circular Clothing Genossenschaft (Cooperative) envisions a zero-waste textile industry, where existing resources are used in a continuous and closed loop. They want to stop the exploitation of the Earth and preserve the planet for future generations by applying the circular economy and Cradle to Cradle® model.[50]

The circular-economy approach for the textile industry aims to eliminate waste and stop the careless use of virgin resources. From the start, garments are designed for reuse, sharing, repair, remanufacturing, recycling, or safe for biological cycles to create a closed-loop system. This

---

[50] "Circular Clothing Is The Future", Circular Clothing, Accessed June 3, 2024, https://circularclothing.org/en/

will minimize the use of resources, and eliminate waste, pollution, and carbon emissions. Currently, less than 1 percent of used clothing is recycled and reused for garments. The Cooperative can take this approach one step further by following the biological or technical cycle of the Cradle to Cradle® model. In this model, valuable and proven solutions are used to design fully circular products: no toxins in, no toxins out.

Unfortunately, small textile labels face great difficulties to becoming circular. So far, only a few large companies with ample resources have been able achieve a certain degree of circularity. The Cooperative will change this. Anyone can join the Cooperative as a member. Start-up fashion labels can join a network to leverage buying power. This access to a certified supply chain ensures the requirements to produce and design Cradle to Cradle Certified ® products.

Within its Start-Up Pro Bono initiative, epeaswitzerland established a functioning textile and fashion supply chain for the Circular Clothing Genossenschaft (Cooperative). The cooperative and its member community share the same ambition and can profit by sharing knowledge and purchasing power in a like-minded community. Products of Circular Clothing Genossenschaft (Cooperative) are Cradle to Cradle Certified® at bronze and gold level.

Being a member of the Circular Clothing Cooperative implies ownership and commitment to a zero-waste textile industry for a better planet. Bundling forces of small fashion and textile labels under the roof of Circular Clothing Cooperative is not limited to the Swiss borders. With the aim of contributing to a paradigm shift in the fashion and textile industry, the Cooperative focuses on Europe. No small label can go circular on its own, but many labels together can make it happen.

Creating Networks to Leverage Circularity

Major multinational companies and SMEs are members of trade associations, which together share common interests. Through relentless and professionally organized lobbying activities, they can influence politics, government, the public, media, and, most importantly, the market. This book has shown how just a few companies in nearly every sector dominate the market. Structuring and building global networks help make them stronger and more influential, for better or worse, depending on their members' objectives. A foundation was established comprising over one hundred multinational companies, all belonging to the network of a large global consulting company, with the objective to pushing the Circular Economy. Regular publication of research reports and case studies on plastics, fashion, packaging, and construction readily accessible to anyone, has made this foundation to a hub for information. It is extremely powerful and influential. Questions arise. Why structure it as a foundation and not an association? What are their real intentions? Is the foundation a smokescreen?

It's called the Ellen McArthur Foundation. A quote from the website: "Business leaders and governments both acknowledge that continued wealth generation requires a new industrial model less dependent on primary energy and materials inputs."[51]

The report "Towards the Circular Economy: Accelerating the scale-up across global supply chains" was produced by the World Economic Forum, the Ellen MacArthur Foundation, and McKinsey & Company. The report aims to reconcile the goal of scaling a circular model with the reality of a global economy and complex multi-tier supply chains. It finds that over US$1 trillion a year could be generated by 2025 for the global economy and one hundred thousand new jobs created over the next five years if companies focused on building circular supply chains to increase the rate of recycling, reuse, and remanufacture.

In 2005, Ellen MacArthur made the record for being the fastest solo sailor to sail around the world. In 2010, she established the Ellen MacArthur Foundation to accelerate the transition to a circular economy. The seventy-one days she spent at sea, carrying everything she needed with her, gave her a new way of looking at the world. She returned with new

---

[51] "Towards the circular economy Vol. 3: accelerating the scale-up across global supply chains", Ellen MacArthur Foundation, Accessed June 3, 2024, https://www.ellenmacarthurfoundation.org/towards-the-circular-economy-vol-3-accelerating-the-scale-up-across-global

insights into the way the world works, as a place of interlocking cycles and finite resources, where the decisions we make today affect what's left for tomorrow.[52]

The Ellen MacArthur Foundation is a non-profit organization that funds original research on the benefits of a circular economy. As an NGO, the Foundation explores opportunities across sectors, and highlights examples of how circular economy principles are being put into practice today. The Foundation supports organizations and individuals with formal learning opportunities through circular economy courses and creates resources for teachers and academics.

Next Generation

Next Generations is a tiny network for "grandchildren-friendly" economic development and economic models based in Switzerland. An open network of individuals and small businesses who want to drive the change towards a sustainable future. Everyone in this network is independent, brings in whatever he/she wants to contribute. A colorful network of a ragtag bunch of knowledge-hungry people who want to work towards a better world. I am a pleased to be a member of Next Generations.

---

[52] "Ellen's story", Ellen MacArthur Foundation, Accessed June 3, 2024, https://www.ellenmacarthurfoundation.org/about-us/ellens-story

CHAPTER 14

Transparency for Consumers - Case Study

Retailers were always very powerful, until online businesses enabled other companies to enter the market through disruptive technologies. Today, online e-commerce is booming, and big players have entered our daily lives, e.g. Amazon, founded 1994. As competition increased among retail and online, the pressure for suppliers to compete for space has expanded to what is "in" and what is "out." It has become a cachet for private label brands to compete against the well-established consumer brands. In the last twenty years, consumer awareness of sustainability has grown substantially, with ethical seals for organic food, animal welfare, fish, fashion, paper, cleaning detergents, cosmetics. For textiles and fashion alone, over 450 eco labels exist, creating an environmentally aware landscape that is hard to miss. Eco-labels offer an identifiable marketing tool to convey a product's environmentally friendly and socially desirable characteristics to consumers. Consumers make choices based on their trust in these labels.

Case Study:

MIGROS creates with M-Check transparency for Near Non-Food Products in cooperation with epeaswitzerland

A large part of the product range from Near Non-Food, offered by the retail chain Migros, has been pre-evalutated for the circular economy. The Migros M-Check five-star pre-evaluation by epeaswitzerland also makes it clear which suppliers can improve their products in terms of circularity.

Migros, Switzerland's largest retailer, is a cooperative (1941) with more than two million members and the largest retail company with a market share of 35 percent and, with over 90,000 employees, the biggest employer in Switzerland. It was founded in 1925 by Gottlieb Duttweiler selling just six basic food products at low prices (coffee, rice, sugar, pasta, coconut fat, soap). A boycott of the brands led to Migros Industries starting to produce their own products. Today, more than twenty thousand products are produced by Migros Industries and sold in the Migros supermarkets in Switzerland or exported to fifty countries. The Migros culture is remarkable: the company does not sell alcoholic beverages nor tobacco (except online), does not pay dividends. One percent of the company's revenue is invested in social and cultural projects.

Migros is expanding its commitment to the circular economy and integrating its supplier network in these efforts. As such, the Near Non-Food range comprising, for example, textiles and household items, is being examined accordingly. These products are pre-evaluated by epeaswitzerland.

On this basis, epeaswitzerland is implementing the M-Check for Migros with regard to two separate dimensions: recyclable ingredients (product circularity) and environmentally friendly ingredients (material health).

In both cases, epeaswitzerland awarded a star rating for the products examined. The top rating of five stars means that the product is fully compliant with the circular economy for recyclable ingredients and that the components have no negative impact on the environment. One-star products, i.e., those made from PVC, still comply with the legal regulations. However, they are not recyclable. Migros is working on ways to produce items based on other materials in the future.

The Migros M-Check pre-evaluation screening from epeaswitzerland is the first step at Migros as the company seeks to ramp up its commitment to the circular economy. After implementation, the five-star rating can be used to find out about product suitability in terms of circular economy credentials. A second step will see Migros additionally include its supplier network based on the five-star rating. Migros will not only recommend that suppliers implement appropriate measures for the creation of a circular economy but also offer active support for suppliers.

Migros drives material health and product circularity innovations for cookware.

In 2023, the European Union considered a proposal to ban widely used, potentially harmful substances known as per- and polyfluoroalkyl substances PFAS or "forever chemicals," which are persistent, toxic, and bioaccumulate in organisms—including people. A large number of US states have limited or banned PFAS in food packaging already. This could become the largest activity of regulation of the chemical industry. PFAS, PFOS, PTFE are used in thousands of products including textiles, medical applications, automotive, flooring, paper, food packaging and nonstick cookware due to their long-term properties in resistance, high temperature, and corrosion.

The threat of banning Forever Chemicals is a slap in the face for companies and industry. The impossible now seems to become reality. The speed of implementation into law is frightening. The industry is not prepared. How does a company get out of this vicious circle?

As part of the Migros Cookware's M-Check pre-evaluation, only one star could be awarded for material health and Circularity because the chemicals used in the coating are excluded from Cradle to Cradle Certified® certification. Consequently, Migros quickly contacted TVS S.p.A. in Italy, the cookware's long-standing supplier, to motivate them to start an innovation project together with epeaswitzerland. After several trials in formulation, sharing of confidential supplier data disclosure, and searching for alternatives, a solution was found that was equivalent to the current product performance. The pre-evaluation achieved a five-star rating for material health and product circularity. This was a great achievement and a good example of creating partnerships initiating product innovations for a better world. By the end of 2023, the products were available for consumers in the Migros stores. Migros announced in their weekly publication, *Migros Magazine*, the success story to a large community of over two million readers.

TVS S.p.A. Made in Italy non-stick cookware has been in operation since 1968. TVS is one of the main international players in the production of cookware in aluminum with a non-stick technology.

THE WORLD HAS ARRIVED AT A CROSSROADS:
EITHER IT GOES CIRCULAR, OR IT VANISHES.

Sustainability is not enough anymore; we must live with what is already on the table. A circular economy makes this possible; a circular economy is necessary.

Nowhere is this as visible as in the global textile industry: It is the world's oldest and biggest industry – and arguably the most polluting. This needs to change, now.

However, there is light at the end of the tunnel: The quest for growth and profit shares common ground with the demand for a circular economy and innovation. Circular innovation creates new products, new business, new chances on the market.

This is neither lofty ideal nor fanciful daydream – this is already being done by pioneering entrepreneurs around the world. All it needs is the will to find new solutions to old problems, the readiness to think outside the box and a reliable network of trust.

→ **ENTREPRENEURS OF THE WORLD, UNITE!**

→ **THE FUTURE OF OUR WORLD NEEDS YOU.**

→ **MAKE YOUR BUSINESS CIRCULAR.**

→ **JOIN THE INNOVATORS.**

→ **BE THE WINNER!**

Epilogue

It's about radicalism, enabling transformation and proving it can be done. It was never my goal to create a large company with hundreds of consultants. I remained loyal to showing a feasible path. I founded this company to create the freedom for the radical implementation of this belief. To guard the freedom of radicalism, it was necessary to create and enable impact to counterbalance the influences of the dominating "linear regime" still in force today. In my opinion, the transition from radical to mainstream is neither a feasible nor desirable role of epeaswitzerland and myself. The signals set are having an impact today, with the circular economy becoming law and projects shining as beacons as proof of concept that Cradle to Cradle® works.

It is up to future generations to multiply it and make it normal in applying true circularity.

Albin Kaelin

References / Companies

A detailed list of companies that are leading the way in transparency and circular practices is listed below. These references serve as case studies and examples for readers interested in the practical application of the concepts discussed in the chapter.

Company	Country	Website	Comments
ALFA Klebstoffe AG	Switzerland	https://alfa.swiss/en/	
Amer Sports Group - Salomon SAS	France	https://www.salomon.com/en-us	
Aquafil S.p.A.	Italy	https://www.aquafil.com/	
Archroma Management GmbH	Switzerland	https://www.archroma.com/	
Archroma Singapore Pte Ltd	Singapore	https://www.archroma.com/	
Artigo Spa	Italy	https://www.artigo.com/	
Asahi Kasei Corp.	Japan	https://www.asahi-kasei.com/company/	
Avient Corporation	United States	https://www.avient.com/	
Bauwerk GROUP SCHWEIZ AG	Switzerland	https://www.bauwerk-parkett.com/int-en	
Bayonix® Stefan Hunger	Germany	https://bayonix.com/	
Braungart EPEA Int. Umweltforschung GmbH	Germany	https://braungart-epea.com/	
Calida AG	Switzerland	https://www.calida.com/en-CH/cms/about/product/100-nature/	

Circular Clothing Genossenschaft	Switzerland	https://circularclothing.org/	
Clariant International Ltd	Switzerland	https://www.clariant.com/en/Corporate	
Climatex AG	Switzerland	https://www.climatex.com/en/	
Cradle to Cradle Product Innovation Institute C2CPII	United States, Netherland	https://c2ccertified.org/	
DyStar Colours Deutschland GmbH	Germany	https://www.dystar.com/	
DyStar Global Holdings (Singapore) Pte. Ltd.	Singapore	https://www.dystar.com/	
EPEA GmbH Part of Drees & Sommer	Germany	https://epea.com/en/	
epeaswitzerland gmbh	Switzerland	https://www.epeaswitzerland.com/	
Fein Elast Group	Austria	https://fein-elast.com/en/	
Grabher Group GmbH	Austria	https://www.grabher-group.company/	
Heierling GmbH	Switzerland	https://www.heierling.ch/en/home	
Inogema GmbH	Germany	https://www.inogema.com/en/?refresh=1687186408	
Karl Dieckhoff GmbH & Co. KG	Germany	http://dieckhoff-textil.de/	
Knauf Ceilings Holding GmbH	Austria	https://www.knaufceilingsolutions.com/en/	
Lanz Natur AG	Switzerland	https://lanur.swiss/?lang=en	
Laufen Bathroom AG	Switzerland	https://www.laufen.com/	

References / Companies

Lauffenmühle GmbH & Co. KG	Germany	n/a	Company in bankruptcy
Lucart SAS	France	https://www.lucartgroup.com/en/	
Luzi AG	Switzerland	https://www.luzi.com/	
Mary Rose GmbH	Austria	https://maryrose.at/en/	
MBDC LLC	United States	https://mbdc.com/	
Migros	Switzerland	https://www.migros.ch/en	
Mondi Group	Austria	https://www.mondigroup.com/en/	
Napapijri	Switzerland	https://www.napapijri.co.uk/circularseries.html	
Next Generation	Switzerland	https://www.next-generations.ch/	
OceanSafe AG	Switzerland	https://www.oceansafe.co/	
Pfister Vorhang Service AG	Switzerland	https://www.pfister.ch/de/services/beratung/vorhang	
PT. Kahatex	Indonesia	http://kahatex-ind.com/sock.html	
Rohner Textil AG	Switzerland	n/a	Company liquidated
SANKO TEKSTİL İŞL.SAN.VE TİC.A.Ş	Türkjye	https://www.sankotextile.com/	
SENS eRecycling	Switzerland	https://www.erecycling.ch/en/	
Siegwerk Druckfarben AG & Co. KGaA	Germany	https://www.siegwerk.com/en/home.html	
Step Zero AG	Switzerland	https://step-zero.com/	
Swarovski	Austria	https://www.swarovski.com/en-CH/	
Tana Chemie GmbH	Germany	https://wmprof.com/se/	

Tanatex Chemicals B.V.	Netherland	https://tanatexchemicals.com/	
Textilcolor AG	Switzerland	https://www.textilcolor.ch/en/	
TRIGEMA Inh. W. Grupp e.K	Germany	https://www.trigema.de/en/sustainability/cradle-to-cradle/	
Trimo d.o.o.	Slovenia	https://www.trimo-group.com/en	
TVS S.p.A.	Italy	https://www.tvs-spa.it/	
VF Corporate	United States	https://www.vfc.com/	
Voegeli AG	Switzerland	https://www.voegeli.ch/de/	
Werner & Mertz Group	Germany	https://werner-mertz.de/en/	
Windmoeller GmbH	Germany	https://windmoeller.de/en/	
Wolford AG	Austria	https://company.wolford.com/	

References / Literature

This section offers a curated list of literature that has influenced the chapter's content. It includes seminal works on sustainability, circular economies, and consumer behavior, providing avenues for further research and learning.

References

A

- Adams, W.M. (2005). *The Future of Sustainability. Re-thinking Environment and Development in the Twenty-first Century*. Published by IUCN Renowned Thinkers Meeting. Retrieved from: www.Iucn.org (16.01.2023)
- Aldoseri, A., Al-Khalifa, K.N., Hamouda, A.M. (2024). *AI-Powered Innovation in Digital Transformation: Key Pillars and Industry Impact*. Engineering Management Program, College of Engineering, Qatar University, Doha P.O. Box 2713, Qatar. https://doi.org/10.3390/su16051790
- Anonymus, 1996 Int. Journal of LCA 1, 119-120 EPEA + Rohner Textil
- Rivière/Soth/Ketelhut, EPEA 1997 from LCA to LCD (Lifecycle Development)

B

- Braungart, M.; Engelfried, J. (1992). *An ‚Intelligent Product System' to Replace ‚Waste Management'*. Fresenius Environmental Bulletin, 1(9), 613-619.
- Braungart, M., Engelfried, J. (1992) *Criteria for Sustainable Development of Products and Production*. Fresenius Environmental Bulletin, 2, 70-77
- Braungart, M., Bondesen, P., Kälin, A., Benson, G. (2008). *Public Goods for Economic Development*. Vienna, Austria: UNIDO Publication. ISBN: 978-92-1-106444-5

- Bosshard. (1997) *Diplomarbeit Die Verbindung von Ökonomie und Ökologie am Beispiel der Rohner Textil AG* (Diplomarbeit). Universität St. Gallen.

C
- Conservation, Odenthal, F. (2018). *Cradle to Cradle. A concept for an ideal circular economy.* Retrieved from: https://www.fairplanet.org/story/cradle-to-cradle-a-concept-for-an-ideal-circular-economy/
- Correira, M., (2022). *Plastikmacher. Die verflucht erfolgreiche Petrochemie*. Le Monde Diplomatique, February 10, 2022. Retrieved from: https://monde-diplomatique.de/artikel/!5826519
- Cox, K. D., Covernton, G.A., Davies, H.L., Dower, J.F., Juanes, F., Dudas, S.E., (2019). *Human Consumption of Microplastics.* Department of Biology, University of Victoria, British Columbia Canada; Hakai Institute, British Columbia Canada; Fisheries and Oceans Canada, British Columbia Canada.

D
- Dubas, D., Angst, D., Leiser, T., Stoll, M., Swiss Federal Council (2022). *2030 Sustainable Development Strategy*. Published by Swiss Federal Council Project management Federal Office for Spatial Development ARE.
- D'Itria, E., Colombi, C. (2022). *Biobased Innovation as a Fashion and Textile Design Must.* Design Department, Politecnico di Milano, 20123 Milan, Italy.
- Deutsche Bundesstiftung Umwelt. (1998). *Stoffstrommanagement -*
- *Herausforderung für eine nachhaltige Entwicklung, Stoffstrommanagement aus der betrieblichen Praxis am Beispiel der Rohner Textil AG*. Sommerakademie 1998.

E
- Elia, V., Gnoni, M.G., Tornese, F. (2017). *Measuring circular economy strategies through index methods: A critical analysis*. Published by Department of Innovation Engineering, University of Salento, Lecce, Italy. Retrieved from: www.elsevier. com/ locate/ jclepr (11.02.2019).

References / Literature

- EIT-Climate KIC (2018). *Digitalisation - unlocking the potential of the circular economy*. Published by EIT-Climate KIC. Retrieved from: https://www.climate-kic.org (16.01.2023)
- Europe Synthesis Report (2023). *Circularity of PET/polyester packaging and textiles in Europe - Synthesis of published research*. Retrieved from: https://www.systemiq.earth/ (12.02.2024)
- European Commission Press release (2022). *European Green Deal: Putting an end to wasteful packaging, boosting reuse and recycling*. Brussel, https://european-union.europa.eu/contact-eu_en
- *EU Strategy for Sustainable and Circular Textiles*. (2022). Communication from the Commission to the European Parliament, The Council, The European Economic and Social Committee and the Committee of the regions. https://eur-lex.europa.eu/legal-content/EN/TXT/HTML/?uri=CELEX:52022DC0141
- EXPO 2000 Hannover. (2000). *Projects around the world of EXPO 2000*. Hannover, Germany: Neue Medien.

F
- Fritjof Capra, Pier Luigi Luisi (2014). *The Systems View of Life. A Unifying Vision*. Cambridge UK, Published by Cambridge University Press.
- Finnveden, G., Arvidsson, R., Björklund, A., Guinée, J., Heijungs, R., Martin, M. (2022). *Six areas of methodological debate on attributional life cycle assessment*. E3S Web of Conference, 349. http://dx.doi.org/10.1051/e3sconf/202234903007
- Frischknecht, R., Rebitzer, G., Spielmann, M., Jungbluth, N., Althaus, H.-J., Doka, G., Nemecek, T. (2005). The ecoinvent Database: Overview and Methodological Framework (7 pp). *The International Journal of Life Cycle Assessment*, 10(1), 3-9. Retrieved from doi: 10.1065/lca2004.10.181.1
- Fernando, J., Reviewer, Brock, T., (2022). *Corporate Social Responsibility (CSR)*. Explained with Examples. Retrieved from: https://www.investopedia.com/terms/c/corp-social-responsibility.asp
- *Facts & Key Figures 2022 of the European Textile and Clothing Industry* (2022). Eurotex - The European Apparel and Textile Confederation. Eurotex Economic and Statistics, Brussel.

G

- Guinée, J.B., Heijungs, R., Huppes, G., Zamagni, A., Masoni, P., Buonamici, R., Ekvall, T., Rydberg, T. (2011). *Life cycle assessment: past, present and future*. Environ. Sci. Technol. 45 (1) (2011) 90-96.
- Gaines, L., Stodolsky, F. (1997). *Life-Cycle Analysis: Uses and Pitfalls*. Conference: Air & Waste Management Association 90th Annual Meeting & Exhibition, Toronto, Canada.
- Guldmann, E., Bocken, N. M. P., and Brezet, H. (2019), *A Design Thinking Framework for Circular Business Model Innovation*, Vol. 7, No. 1, pp. 39-70
- Global Petrochemicals New Build and Expansion Projects Outlook, 2023-2027 (2023) published by: GlobalData Plc Intelligence Center. Rretrieved from: https://www.globaldata.com/
- Gupta, S., Rhyner, J. (2022). *Mindful Application of Digitalization for Sustainable Development: The Digitainability Assessment Framework*. Published in: Sustainability 2022, 14, 3114. https://doi.org/10.3390/su14053114
- Gorman, M. E., Mehalik, M.M (1998). *Towards a Sustainable Tomorrow:*
- *Three Cases and a Moral. Consumption in the Global Environment of the 21st Century*. eds. Werhane, P.H., Westra, L.
- Gorman, M. E., Mehalik, M.M., Werhane, P.H. (1999). *Ethical and Environmental Challenges to Engineering*. New York, Upper Saddle River, Prentice Hall.
- Gorman, M.E., Werhane, P.H., Mehalik, M.M. Designtex, Inc. (A). Darden Case No. UVA-E-0099, Available at SSRN: https://ssrn.com/abstract=908141 or http://dx.doi.org/10.2139/ssrn.908141
- Gorman, M.E. (1999). *Transforming Nature: Ethics, Invention and Discovery*.
- International Journal of Technology and Design Education, vol.9, 195–196.

H

- Heim, H., Hopper, C. (2022). *Dress code: the digital transformation of the circular fashion supply chain*. International Journal of Fashion Design, Technology and Education, 15:2, 233-244, DOI: 10.1080/17543266.2021.2013956. To link to this article: https://doi.org/10.1080/17543266.2021.2013956

References / Literature

- Hawken, P., Lovins, A., Hunter-Lovins, L. (2000). *Natural Capitalism: Creating the Next Industrial Revolution*. London, England: Earth Scan Publications Ltd.. ISBN: 1853834610
- Herkströter, R., Kälin, A., Krajner, M. (2016). *Cradle to Cradle® - Parquet for Generations Respect for Resources and Preservation for Future*. St. Margrethen, Switzerland: Bauwerk Parkett AG, Bäch, Switzerland: EPEA Switzerland GmbH.
- https://www.unep.org/resources/report/chemicals-plastics-technical-report - :~:text=Based on the latest studies, a wide range of applications. (24.10.2023)
- https://saicmknowledge.org/chemicals-management-toolkit-toy-sector (24.10.2023)
- https://www.globaldata.com/store/report/petrochemicals-new-build-and-expansion-projects-market-analysis/ (24.10.2023)
- https://www.foodpackagingforum.org/news/report-highlights-potential-hazards-of-plastic-bottle-supply-chain (24.10.2023)
- https://www.foodpackagingforum.org/news/plastics-identified-as-source-of-legal-liability (24.10.2023)
- https://www.theguardian.com/environment/2023/nov/08/plastic-waste-spiralling-out-of-control-across-africa-analysis-shows?CMP=Share_AndroidApp_Other (14.11.2023)
- https://www.madetrade.com/blogs/made-trade-magazine/what-is-regenerative-fashion
- https://netimpact.org/blog/regenerative-systems-can-change-fashion-industry
- https://sustainablebrands.com/read/product-service-design-innovation/the-shift-from-sustainable-to-regenerative-design. PAUL C. HUTTON.
- https://swissrecycle.ch/de/wertstoffe-wissen/studiengaenge (24.10.2023)
- https://www.circulareconomyforum.at/european-green-deal-time line/https://www.econstor.eu/bitstream/10419/248441/1/1782081 615.pdf
- https://www.bildung-schweiz.ch/angebote/
- https://www.studieren-studium.com/master/studieren/nachhaltigkeit-Schweiz
- https://www.zhaw.ch/de/sml/weiterbildung/detail/kurs/cas-managing-circular-economy/?pk_campaign=sml_A-wb-

- cas_K-mancirceco_O-gm-imm_N-cas-managing-circular-economy&pk_kwd=MK-sea_MT-txt_P-google-ch_MF-txt_T-dech_G-126680345427_C-kreislaufwirtschaft%20%2Bcas_Z-cpc&gclid=Cj0KCQiAz9ieBhCIARIsACB0oGJH37HycEJrYimZoBnNYo4yTIZtIc9B_dXnmAugoEUtUipBs0HWmbIaArFfEALw_wcB (29.01.2023)
- https://online.professionalprogramsmit.com/circular-economy?utm_campaign=mpe-cie-eng&utm_source=ppc&utm_medium=adwords&utm_content=mpe-cie-eng-gads-seaeuropa&utm_term=courses%20in%20sustainability&utm_location=1003141&utm_network=g&gclid=Cj0KCQiAz9ieBhCIARIsACB0oGJQQzpG48iFu_5rim_qi4SC2OZfYxp82KxkgQ6NcZl5TScetBFTVmAaAuVhEALw_wcB (29.01.2023)
- https://www.epfl.ch/education/continuing-education/key-actors/iml/certificate-advanced-studies/ (29.01.2023)
- https://www.bfh.ch/de/studium/master/circular-innovation-and-sustainability/?gclid=Cj0KCQiAz9ieBhCIARIsACB0oGK4T-3GTXwSUwtjw3hNxqeZwcyBn3hqz_-ujNDeMWlY-ch5Hs42glAaAumuEALw_wcB (29.01.2023)

I

- International Living Future Institute (ILFI): https://living-future.org/

J

- Jamieson, D. (2001). *A Companion to Environmental Philosophy*. Oxford UK, published by Blackwell Publishers Ltd.

K

- Keiner, M. (2005). History, definition(s) and models of sustainable development. ETH Zürich. Retrieved from: https://doi.org/10.3929/ethz-a-004995678 (16.01.2023)
- Korhonen, J., Nuur, C., Feldmann, A., Eshetu Birkie, S. (2017). *Circular economy as an essentially contested concept*. Retrieved from: www.elsevier.com/locate/jclepro (11.02.2019)
- Kopnina, H. (2018). Circular economy and Cradle to Cradle in educational practice. *Journal of integrative Environmental*

Sciences, 15 (1), 119-134. https://doi.org/10.1080/194381
5X.2018.1471724
- Krithivasan, R., Belliveau, M., Lani, A. (2023). Hidden Hazards: The chemical footprint of a plastic bottle. Published by: DefendOurHealth.org. Portland, Maine.
- Kälin, A., Rivière, A., Ketelhut, R., Braungart, M. (2002). *From Ecoefficiency to Overall Sustainability in Enterprises*. Climatex® LifeguardFR™ Upholstery Fabrics - The Chronicle of a Sustainable Product Redesign.
- Kälin, A., Mehalik, M.M. (2000). *The Development of Climatex®Lifecycle™, a Compostable, Environmentally Sound Upholstery Fabric*. Sustainable Solutions: Sustainable and Eco-Product and Service Development. Ed. Martin Charter. London, England: Greenleaf Publishing.

L

- Latour, B. (2019). *Down to Earth. Politics in the New Climatic Regime*. Paris. First published by Éditions La Découverte; Reprinted Cambridge UK by Polity Press.
- Life Cycle Initiative, hosted by UNEP. *Social Life Cycle Assessment (S-LCA)*. Retrieved from: https://www.lifecycleinitiative.org/starting-life-cycle-thinking/life-cycle-approaches/social-lca/ (20.04.2022)

M

- McDonough, W., Braungart, M. (2002). *Cradle to Cradle. Remaking the Way We Make Things*. New York, published by North Point Press.
- McDonough, W., Braungart, M. (2013). *The Upcycle: Beyond Sustainability - Designing for Abundance*. New York, published by North Point Press.
- McDonough, W., Braungart, M. (2008). *Die nächste industrielle Revolution. Die Cradle to Cradle-Community*. Hamburg, published by Europäische Verlagsanstalt.
- Michaels, F.S. (2011). *Monoculture. How One Story is Changing Everything*. Canada. Published by Red Clover Press.
- McElroy, M.W., Jorna, R.J., van Engelen, J. (2007). *Sustainability Quotients and the Social Footprint*. Published by John Wiley and

Sons Ltd & European Research Press Ltd. Retrieved from: https://onlinelibrary.wiley.com/doi/abs/10.1002/csr.164 (03.05.2023)
- McDonough, W. (1993). *A Boat for Thoreau. Architecture, Ethics and the Making of Things*. Business Ethics: The Magazine of Corporate Responsibility, 7(3), 26-29.
- McDonough, W. (1993). *A Centennial Sermon, Design Ecology Ethics and the Making of things*. Designtex 1993, Environmentally Intelligent Textiles 1995
- Mehalik, M.M. (2000). *Technical and Design Tools: The Integration of ISO 14001, Life Cycle Development, Environmental Design and Cost Accounting*. ISO 14001 Case Studies and Practical Experience. Ed Ruth Hillary. London, England: Greenleaf Publishing.
- Mehalik, M. M. (2000). *Sustainable Network Design: A Commercial Fabric Case Study*. Interface 30(3), 180-189. DOI:10.1287/inte.30.3.180.11659
- Mayer-Ries, J.F. et al. (1998). *Zwischen Globalen und Lokalen Interessen* (Kälin, A., Produkt Redesign - der Natur zuliebe am Beispiel eines kompostierbaren, umweltverträglichen Möbelbezugstoffes). Rehburg-Loccum, Germany: Evangelische Akademie Loccum. ISBN: 978-3817216987

N
- Nebel, B. (2020). *Cradle to Cradle, LCA and Circular Economy: a love triangle*. Published NZ Manufacturer magazine. Retrieved from: https://www.thinkstep-anz.com/resrc/blogs/cradle-to-cradle-life-cycle-assessment-and-circular-economy-a-love-triangle/

O
- Orr, D.W. (2004). *Earth in Mind. On Education, Environment and the Human Prospect*. Washington, DC, published by First Island Press.

P
- Porrit, J., (2020). *Hope in Hell. A Decade to confront the Climate Emergency*. London, published by Simon & Schuster.
- Princen, T. (2010). *Treading Softly. Paths to Ecological Order*. Cambridge Massachusetts, published by Massachusetts Institute of Technology.

R

- Reikea, D., Vermeulen, W.J.V., Witjes, S. (2017). *The circular economy: New or Refurbished as CE 3.0? – Exploring Controversies in the Conceptualization of the Circular Economy through a Focus on History and Resource Value Retention Options.* Retrieved from: https://www.sciencedirect.com/science/article/pii/S0921344917302756?via%3Dihub
- Report Federal Council (2023). *Waste management, waste avoidance, waste planning, measurement.* (in fulfillment of the postulates Bourgeois 20.3062, Munz 20.3090, Clivaz 20.3727, Gapany 20.4411, Chevalley 20.3110 and the Commission for the Environment, Spatial Planning and Energy of the National Council.
- Riess (1998). *Rohner Textil Massgeschneidertes Beurteilen. Öko-Controlling.* Zürich, Schweiz: Bulletin ETH Zürich, 268/Jan 98.
- Rytec (2016). *Geschäftsmodelle zur Förderung einer Kreislaufwirtschaft. Grundlagenbericht und Workshopergebnisse.* Biel, Schweiz: sanu durabilitas. vol.2.

S

- Spangenberg, J.H. (1998). *Sustainability Indicators – a Compass on the Road Towards Sustainability.* Published by Sustainable Europe Research Institute SERI Germany. Retrieved from: https://www.researchgate.net/publication/254460859 (16.01.2023)
- Santa-Maria, T., Vermeulen, W.J.V., Baumgartner, R. J. (2020). *Framing and assessing the emergent field of business model innovation for the circular economy: A combined literature review and multiple case study approach.* Retrieved from: www.elsevier.com/locate/spc (22.04.2022)
- Sandford, C. (2021, January). *The Regenerative Business: Redesign Work & Cultivating Human Potential.* Carol Sandford Institute. Regenerative Business Education.
- Retrieved from: https://growensemble.com/regenerative-business/
- Spörri et al. (2021): *Die Hürden gegen Ressourceneffizienz und Kreislaufwirtschaft abbauen.* Studie zum gleichnamigen Postulat 18.3509 von Ständerat Ruedi Noser. Schlussbericht im Auftrag des Bundesamts für Umwelt. EBP Schweiz AG, Berner Fachhochschule.

- Schweizerische Eidgenossenschaft, Federal Office for the Environment FOEN. Circular Economy. Retrieved from: https://www.bafu.admin.ch/bafu/de/home/themen/wirtschaft konsum/fachinformationen/kreislaufwirtschaft.html
- Stecher, B. (2020). *A vision of how to move towards a regenerative fashion system.* Blog, Biomimicry In Design) https://biomimicry.org/a-vision-of-how-to-move-towards-a-regenerative-fashion-system/

T

- Todorov, V., Marinova, D. (2009). *Models of Sustainability.* Published by University of Forestry Sofia & Curtin University of Technology Western Australia. Retrieved from: https://www.researchgate.net/publication/228874179 (18.05.2022)
- Takacs, F., Stechow, R., Frankenberger, K. (2020). *Circular Ecosystems: Business Model Innovation for the Circular Economy.* White Paper of the Institute of Management & Strategy, University of St. Gallen.
- The Ministry of Infrastructure and the Environment, Ministry of Economic Affairs, Ministry of Foreign Affairs, Ministry of Interior and Kingdom Relations (2016). *A Circular Economy in the Netherlands by 2050. Government-wide Programme for a Circular Economy.* Retrieved from: www.government.nl/circular-economy (17.02.2019).
- The Intergovernmental Panel on Climate Change. (2023). *Working group I: Climate Change 2021: The Physical Science Basis.* Retrieved from: https://www.ipcc.ch/report/sixth-assessment-report-working-group-i/
- The Intergovernmental Panel on Climate Change. (2023). *Working group, II Climate Change 2022: Impacts, Adaptation and Vulnerability.* Retrieved from: https://www.ipcc.ch/report/ar6/wg2/
- The Intergovernmental Panel on Climate Change. (2023). *Working group, III Climate Change 2022: Mitigation of Climate Change.* Retrieved from: https://www.ipcc.ch/report/ar6/wg3/
- The Intergovernmental Panel on Climate Change. (2023). *Special Report on the Ocean and Cryosphere in a Changing Climate.* Retrieved from: https://www.ipcc.ch/srocc/

- The Intergovernmental Panel on Climate Change. (2023). *Global Warming of 1.5 °C*. Retrieved from: https://www.ipcc.ch/sr15/
- The Intergovernmental Panel on Climate Change. (2023). *Climate Change and Land*. Retrieved from: https://www.ipcc.ch/srccl/
- The Intergovernmental Panel on Climate Change. (2023). *Key findings in AR6*. Retrieved from: https://climate.selectra.com/en/news/ipcc-report-2022
- Tobias Stucki, Martin Wörter (2021). *Statusbericht der Schweizer Kreislaufwirtschaft*. Erste repräsentative Studie zur Umsetzung der Kreislaufwirtschaft auf Unternehmensebene. Schlussbericht im Auftrag des Bundesamts für Umwelt & Circular Economy Switzerland. Berner Fachhochschule Wirtschaft, ETH Zürich, KOF Konjunkturforschungsstelle.
- The Federal Assembly, The Swiss Parliament, Clivaz, C., Jauslin, M.S. (2020). *Strengthen the Swiss circular economy*. 20,433 Parliamentary Initiative (2020). Retrieved from: https://www.parlament.ch/de/ratsbetrieb/suche-curia-vista/geschaeft?AffairId=20200433
- Treude, S. (2022). European economic policy and the European Green Deal: an institutionalized analysis. Scientific publications from the department Economics, No. 35-2022, Koblenz University of Applied Sciences, Department of Economics, Koblenz
- Tatari, A., Jinaru, A., Dorbota, G., Niculescu, G. (2021). *Horizons for sustainability „Constantin Brâncuşi"*. University of Târgu-Jiu, Issue 2/2021
- Thompson, A.A., Strickland, A.J. (2001). *Strategic Management Concept + Cases*. McGraw-Hill/Irvin, Pennsylvania State University (2001), ISBN: 0072314990, 9780072314991

U
- United Nations Environment Programme, Belgian Federal Public Planning Service Sustainable Development (2009). Guidelines for Social Life Cycle Assessment of Products. Retrieved from: www.unep.org (20.04.2022)
- Umweltmanagementsysteme in der Textilindustrie Rohner Textil AG. Kissing, Germany: WEKA MEDIA GmbH&Co.KG

V

- Vigne, S., Mason, L. (2022). *All eyes turn to the European Court of Human Rights to assess future of rights-based climate litigation.* Universal Rights Group, February 1. Retrieved from: https://www.universal-rights.org/blog/all-eyes-turn-to-the-european-court-of-human-rights-to-assess-future-of-rights-based-climate-litigation/ (16.01.2023)
- Van der Velden, M. (2018). *Digitalization and the UN Sustainable development Goals: What role for design.* Department of Informatics University of Oslo. Interaction Design and Architecture(s) Journal - IxD&A, N.37, 2018, pp. 160-174
- Von Abendroth, G. (2008). *Gemacht für die Zukunft: Kreislaufwirtschaft in der Unternehmenspraxis.* Hamburg, Deutschland: Murmann Publishers. ISBN-13: 978-3867740364
- Vinuesa, R., Azizpour, H., Leite, I., Balaam, M., Dignum, v., Domisch, S., Felländer, A., Langhans, S.D., Tegmark, M., Nerini, F.F. (2020). *The role of artificial intelligence in achieving the Sustainable Development Goals.* Nature Communications (2020) 11:233. https://doi.org/10.1038/s41467-019-14108-y
- Von Weizsäcker, E.U., Stigson, B., Seiler-Hausmann, J.D. et al. (2001). *Von Ökoeffizienz zu nachhaltiger Entwicklung in Unternehmen. From Eco-Efficiency to Overall Sustainable Development in Enterprises.* (Kälin, A. Positiv definierter Chemikalieneinsatz als Voraussetzung für die Schliessung von Material- und Wasserkreisläufen: Das Beispiel des Möbelbezugstoffes Climatex®Lifecycle™ der Rohner Textil AG. Wuppertal, Germany: Wuppertal Institut f. Klima, Umwelt, Energie. ISBN 978-3800623013
- Von Weizsäcker, E.U., Seiler-Hausmann, J-D. (1999) *Ökoeffizienz. Management der Zukunft.* Basel, Schweiz: Birkhäuser Verlag ISBN-13: 9783764360696

W

- William McDonough & Partners (1992). *The Hannover Principles. Design for Sustainability. Prepared for EXPO 2000 The World's Fair Hannover Germany.* Published by William McDonough Architects.
- World Economic Forum (2019). *A New Circular Vision for Electronics. Time for a Global Reboot. In support of the United*

References / Literature

Nations E-waste Coalition. Retrieved from: www.weforum.org (17.02.2019).
- Williams, N., Collet, C. (2020). *Biodesign and the Allure of "Grow-made" Textiles. An Interview with Carole Collet.* published by GeoHumanities. https://doi.org/10.1080/2373566X.2020.1816141
- Williams, D., Forst, L., Vladimirova, D., Evans, S. (2021). *Building our worlds: co-developing future scenarios as a methodology for fashion researchers and designer-innovator***s**. Safe Harbours for Design Research, 14th EAD Conference, October 2021.
- Wahl, D.C. (2016). *Designing Regenerative Cultures*. Axminster England. Published by Triarchy Press.
- Winter, G. (1998). *Das umweltbewusste Unternehmen: Die Zukunft beginnt heute, Kapitel: Umweltmanagement in Kleinunternehmen. Beispiel Rohner Textil AG*. C.H. München, Deutschland: Beck Verlag; 6. Edition. ISBN-13: 978-3800623013
- Werhane, P.H. (1999). *Moral Imagination and Managing Decision Making*. Oxford, England: Oxford University Press.

Trademarks

Any trademarks or proprietary terms used throughout the chapter is duly acknowledged, ensuring respect for intellectual property and encouraging ethical consumption.

About the Author

Albin Kälin, CEO, is a Swiss citizen and the first ever to implement a Cradle to Cradle® product worldwide in 1992; the award winning Climatex®, and to turn around a Swiss textile mill to become profitable and surpass all environmental regulations in the process. In 2005-2009, Prof Michael Braungart, one of the two founders of Cradle to Cradle®, appointed Albin Kälin as CEO of EPEA Internationale Umwetltforschung GmbH in Hamburg, Germany, and in 2008 as CEO of EPEA Nederland b.v. to implement successfully Cradle to Cradle® concepts and products within industry and to build EPEA's Strategy for the future.

Environmental Protection Encouragement Agency, epeaswitzerland gmbh,, was founded in 2009. It is 100 percent owned by Albin Kaelin GmbH to support businesses globally, to innovate, and become safe and circular. Cradle to Cradle® projects are implemented with a team of international professional industry and business managers. It is a different way of thinking; this is why it is so challenging. We need many lighthouses who can prove the different way of thinking is a success. How can we reach the tipping point soon?

Website: www.epeaswitzerland.com.

Social Media: epeaswitzerland

www.ingramcontent.com/pod-product-compliance
Lightning Source LLC
Chambersburg PA
CBHW052136070526
44585CB00017B/1851